Lecture Notes in Computer Science 8302

Commenced Publication in 1973
Founding and Former Series Editors:
Gerhard Goos, Juris Hartmanis, and Jan van Leeuwen

Vasudha Bhatnagar   Srinath Srinivasa (Eds.)

Big Data Analytics

Second International Conference, BDA 2013
Mysore, India, December 16-18, 2013
Proceedings

Springer

Volume Editor

Vasudha Bhatnagar
South Asian University
Department of Computer Science
Akbar Bhavan, Chanakyapuri
New Delhi, India
E-mail: vbhatnagar@cs.sau.ac.in

Srinath Srinivasa
International Institute
of Information Technology
Banglore, India
E-mail: sri@iiitb.ac.in

ISSN 0302-9743   e-ISSN 1611-3349
ISBN 978-3-319-03688-5   e-ISBN 978-3-319-03689-2
DOI 10.1007/978-3-319-03689-2
Springer Cham Heidelberg New York Dordrecht London

Library of Congress Control Number: 2013953235

CR Subject Classification (1998): H.3, H.2.8, H.2, I.2, H.4, I.5, F.2, G.2, H.5

LNCS Sublibrary: SL 3 – Information Systems and Application
incl. Internet/Web and HCI

Typesetting: Camera-ready by author, data conversion by Scientific Publishing Services, Chennai, India

Printed on acid-free paper

Springer is part of Springer Science+Business Media (www.springer.com)

Preface

Contemporary digital world comprises text, images, videos, and multiple forms of semi-structured data that are inter-linked and inter-related in complex networks. Pervasive in both commercial and scientific domains, these data present innumerable opportunities for discovering patterns, accompanied by challenges of matching magnitude. The data deluge has fueled the creativity of data-curious researchers and has led to the rapid emergence of new technologies in data analytics. The major impetus has come from the variety, variability, and veracity of data in addition to their infinitely growing volume (the 4 V's of Big Data).

The second edition of the International Conference on Big Data Analytics (BDA 2013) was held during December 16–18, 2013, in Mysore, India, to congregate researchers, practitioners and policy makers, for sharing their works and experiences in development of methods and algorithms for big data analytics. The conference attracted 49 submissions in all, of which 45 were submitted for the research track and four for the industry track. All submitted papers were subjected to plagiarism check before review. Each paper was reviewed by at-least three reviewers and the review comments were communicated to the authors. The review process resulted in the acceptance of nine regular papers and one short paper for the research track. One industry paper was accepted, leading to an overall acceptance rate of 22%. This volume includes the accepted papers, tutorials, and invited papers presented during the conference.

The section "Mining Social Media Data" comprises four papers. The tutorial article by Mehta and Subramaniam focuses on methods for doing entity analytics and integration using large twitter data sets. They review the state of the art, and present new ideas on handling common research problems such as event detection from social media, summarization, location inference and fusing external data sources with social data. Sureka and Agrawal address the problem of detecting copyright infringement of music videos on YouTube. They propose an algorithm that mines both video and uploader meta-data, and uses a rule-based classifier to predict the category (original or copyright-violated) of the video. Jain et al. investigate user behavior based on the temporal dimension of tweets and relate it to the evolution of topics on Twitter OSN. Based on a novel metric called "tweet strength," topics are identified along with the users driving the evolution. Bhargava et al. study the problem of authorship attribution of tweets, for forensic purposes. The proposed method extracts stylometric information from the collected data set to predict authors, using classification algorithms.

The section "Perspectives on Big Data Analytics," which comprises three papers, opens with a tutorial paper by Lakshminarayan. The paper presents some fundamental methods of dimensionality reduction and elaborates on the main algorithms. The author also points to next-generation methods that seek to identify structures within high-dimensional data, not captured by second-order statistics. The invited paper by Mondal discusses the role of crowd-driven data collection in big data analytics and opportunities presented by such collections. Kiran addresses the intermittance problem in large transaction databases. The paper introduces quasi-periodic-frequent patterns, which provide useful information and are immune to intermittance problem.

The section "Graph Analytics" consists of papers related to mining of large graphs. Das and Chakravarthy present a survey of graph algorithms and identify the challenges of adapting/extending algorithms for the analysis of large graphs using the Map-Reduce programming model. Tripathy et al. study the characteristics of complex networks in the game of cricket, where dyadic relationships exist among a group of players. Properties such as average degree, average strength, and average clustering coefficients are found to be directly related to the performances of the teams. Parveen and Nair propose techniques for effective and efficient visualization of small-world networks in a similarity space. An algorithm for the *visual assessment of cluster tendency* is presented for efficient hierarchical graphical representation of large networks.

The section "Practice of Big Data Analytics" consists of three papers describing practical applications. Elisabeth et al. present a tourist recommender system using GPS data collected from rental tourist cars. Misra et al. present a case study to demonstrate the performance advantage of Hadoop-based ETL tools over the traditional tools. Lakshminarayan and Baron investigate and report on the of application of big data analytics in manufacturing of integrated circuits.

We gratefully acknowledge the support extended by the University of Delhi and the University of Aizu. We owe gratitude to MYRA School of Business in Mysore for organizing the conference and extending their hospitality. Thanks are also due to our sponsors: E-Bay and IBM India Research Lab. We also thank all the Program Committee members and external reviewers for their time and diligent reviews. Ramesh Agrawal performed a plagiarism check on submissions; thanks to him and his team. The Organization Committee and student volunteers of BDA 2013 deserve special mention for their support. Special thanks to the Steering Committee members. Finally, thanks to EasyChair for making our task of generating this volume smooth and simple.

December 2013

Vasudha Bhatnagar
Srinath Srinivasa

Organization

Steering Committee

R.K. Arora	IIT Delhi, Delhi, India
Subhash Bhalla	University of Aizu, Japan
Sharma Chakravarthy	University of Texas at Arlington, USA
Rattan Datta	Indian Meteorological Department, Delhi, India
S.K. Gupta	IIT, Delhi, India (Chair)
H.V. Jagadish	University of Michigan, USA
D. Janakiram	IIT Madras, India
N. Vijayaditya	Government of India

Executive Committee

General Chair

D. Janakiram   IIT Madras, India

Program Co-chairs

Srinath Srinivasa   IIIT, Banglore, India
Vasudha Bhatnagar   University of Delhi, India

Organizing Chair

Shalini Urs   ISiM, Mysore, India

Publicity Chair

Vikram Goyal   IIIT, Delhi, India

Proceedings Chairs

Subhash Bhalla   University of Aizu, Japan
Naveen Kumar   University of Delhi, India

Industry Chair

Vijay Srinivas Agneeswaran   Impetus Labs, India

Tutorials Chair

Jaideep Srivastava   University of Minnesota, USA

PhD Symposium Chair

Maya Ramanath   IIT Delhi, India

Local Organizing Committee

Abhinanda Sarkar   MYRA School of Business, India
Naveen Kumar   University of Delhi, India
Subhash Bhalla   University of Aizu, Japan

Program Committee

Vijay Srinivas Agneeswaran   Impetus Labs, Bangalore, India
Ramesh Agrawal   Jawaharlal Nehru University, New Delhi, India
Avishek Anand   Max Planck Institute, Germany
Amitabha Bagchi   Indian Institute of Technology, Delhi, India
Srikanta Bedathur   Indraprastha Institute of Information
                                     Technology (IIIT), Delhi, India
Subhash Bhalla   University of Aizu, Japan
Raj Bhatnagar   University of Cincinnati, USA
Arnab Bhattacharya   Indian Institute of Technology, Kanpur, India
Indrajit Bhattacharya   IBM Research, India
Gao Cong   Nanyang Technological University, Singapore
Prasad Deshpande   IBM Research, India
Lipika Dey   TCS Innovation Labs, Delhi, India
Dejing Dou   University of Oregon, USA
Haimonti Dutta   Columbia University, USA
Shady Elbassuoni   American University, Beirut, Lebanon
Rajeev Gupta   IBM Research, India
Sharanjit Kaur   University of Delhi, India
Akhil Kumar   Penn State University, USA
Naveen Kumar   University of Delhi, USA
Choudur Lakshminarayan   Hewlett-Packard Laboratories, USA
Ulf Leser   Institut für Informatik, Humboldt-Universität
                                     zu Berlin, Germany
Ravi Madipadaga   Carl Zeiss, India
Sameep Mehta   IBM Research, India
Mukesh Mohania   IBM Research, India
Yasuhiko Morimoto   Hiroshima University, Japan
Joydeb Mukherjee   Impetus Labs, India
Saikat Mukherjee   Siemens, India
Mandar Mutalikdesai   Siemens, India
Felix Naumann   Hasso-Plattner-Institut, Potsdam, Germany
Hariprasad Nellitheertha   Intel, India
Anjaneyulu Pasala   Infosys Labs, India

Adrian Paschke	Freie Universität Berlin, Germany
Jyoti Pawar	Goa University, India
Lukas Pichl	International Christian University, Japan
Krishna Reddy Polepalli	International Institute of Information Technology, Hyderabad, India
Kompalli Pramod	International Institute of Information Technology, Hyderabad, India
Mangsuli Purnaprajna	Honeywell, India
Sriram Raghavan	IBM Research, India
S. Rajagopalan	International Institute of Information Technology, Bangalore, India
Muttukrishnan Rajarajan	City University
Raman Ramakrishnan	Honeywell, India
Chandrashekar Ramanathan	International Institute of Information Technology, Bangalore, India
Markus Schaal	University College Dublin, Ireland
Srinivasan Sengamedu	Komli Labs, Bangalore, India
Shubhashis Sengupta	Accenture, India
Mark Sifer	University of Wollongong, New Zealand
Jaideep Srivastava	University of Minnesota, USA
Shamik Sural	Indian Institute of Technology, Kharagpur, India
Ashish Sureka	Indraprastha Institute of Information Technology (IIIT), Delhi, India
Asoke Talukder	Interpretomics Labs, Bangalore, India
Srikanta Tirthapura	Iowa State University, USA
Sunil Tulasidasan	Los Alamos National Laboratories, USA
Sujatha Upadhyaya	Independent Consultant, Bangalore, India
Shalini Urs	International School of Information Management, Mysore, India

Additional Reviewers

Adhikari, Animesh	P, Deepak
Agarwal, Manoj	Prateek, Satya
Burgoon, Erin	Puri, Charu
Correa, Denzil	Rachakonda, Aditya
Gupta, Shikha	Ranu, Sayan
Jog, Chinmay	Ravindra, Padmashree
Kulkarni, Sumant	Sreevalsan-Nair, Jaya
Lal, Sangeeta	Telang, Aditya

Table of Contents

Big Data in Practice

Tutorial : Social Media Analytics

Sameep Mehta and L.V. Subramaniam

IBM India Research Lab, New Delhi, India
{sameepmehta,lvsubram}@in.ibm.com

Abstract. In this tutorial we present an overview of some of the common tasks in analyzing text messages on Social Media (mostly on microblogging sites). We review the state of the art as well as present new ideas on handling common research problems like Event Detection from Social Media, Summarization, Location Inference and fusing external data sources with social data. The tutorial would assume basic knowledge of Data Mining, Text Analytics and NLP Methods.

1 Introduction

Social networking sites like Twitter and Facebook have proven to be popular outlets for information dissemination during crises. It has been observed that information related to crises is released on social media sites before traditional news sites [25], [36]. During the Arab Spring movement, Twitter was used as an information source to coordinate protests and to bring awareness to the atrocities [34]. In recent world events, social media data has been shown to be effective in reporting earthquakes [46], detecting rumors [40] and identifying characteristics of information propagation [45]. Some of the key properties of Social Media data which require special handling are:

Volume: Approximately 400 million tweets are now posted on Twitter every day [50]. The volume is continuously increasing. Event detection methods need to be scalable to handle this high volume of tweets.

Velocity: The underlying events in tweets can quickly evolve with the real-world issues. Events are naturally temporal and an event could be characterized by the rate at which the tweets of the event arrive.

Informal Use of Language: Twitter users produce and consume information in a very informal manner compared with traditional media. Mis-spellings, abbreviations, and slang are rampant in tweets, which is exacerbated by the length restriction (a tweet can have no more than 140 characters).

In this tutorial we focus on a subset of research problem in Social Media around Event Detection, Location Inference, Summarization and fusion with external information sources.

2 Event Detection

Event detection in traditional media is also known as Topic Detection and Tracking (TDT). A pilot study on this task was performed in a year-long study by

V. Bhatnagar and S. Srinivasa (Eds.): BDA 2013, LNCS 8302, pp. 1–21, 2013.

Allan et al. [23]. Yang et al. [54] modeled news articles as documents to detect topics. They transformed the tweets into vector space using the TF-IDF representation and evaluated two clustering approaches: Group-Average Agglomerative Clustering (GAAC) for retrospective event detection, and Incremental Clustering for new event detection. They discovered that the task of new event detection was harder than that of detecting events retrospectively. In [24], the authors focused on the problem of online event detection. The authors approached the problem as a document-query matching problem. A query was constructed using the k most frequent words in a story. If a new document did not trigger existing queries, the document was considered to be part of a new event. In [32], the authors addressed the problem of detecting *hot* bursty events. They introduced a new parameter-free clustering approach called feature-pivot clustering, which attempted to detect and cluster bursty features to detect hot stories.

Kumaran et al. [37] formulated the task of event detection as a classification problem. Each document could be classified into two classes: a new story or an old one. A story is represented as a set of features constructed by evaluating the cosine similarity of the document with a story that had the highest similarity in terms of words, topic words, and/or named entities. The authors observed that RBF kernel had the best classification accuracy on a dataset consisting of news articles. Named entities such as people, location, organization, and date/time were used explicitly in [55] as a means to enhance new event detection. A graph-based hierarchical approach was proposed to solve the problem. New documents were compared to the representative document in a cluster and comparisons were made only to individual documents of a small number of clusters to improve the efficiency of their approach.

An attempt to detect earthquakes using Twitter users as social sensors was carried out by Sakaki et al. in [46]. The temporal aspect of an event was modeled as an exponential distribution, and the probability of the event was determined based on the likelihood of each sensor being incorrect. A spatio-temporal model was learned from classified training samples, and the authors used SVM with a linear kernel to evaluate the effectiveness of their approach. In [47], the authors converted a series of blog posts into a keyword graph, where nodes represented words and links represented co-occurrence. On this graph, the authors applied community detection methods to detect communities of related words or events. Becker et al [26], tackled event detection in Flickr. The authors leveraged the meta data of shared images to create both textual and non-textual features and proposed the use of individual distance measures for each feature. These features were used to create independent partitions of the data and finally the partitions were combined using a weighted ensemble scheme to detect event clusters. Weng and Lee [52] constructed word signals using Wavelet Transformation to detect events in tweets. The authors used a modularity-based graph partitioning approach on a correlation matrix to get event clusters. One of the concerns with this approach is the use of TF-IDF based word representation, which can be expensive due to the maintenance required for a evolving vocabulary in

Twitter. Additionally, the use of modularity based clustering approach reduces the scalability of the approach on high volume Twitter streams.

Few existing approaches are designed to tackle the problem of event detection in streaming Twitter data. The current state-of-the-art for the detection of events in a stream was presented in [43]. The authors recognized the need for faster approaches to tackle the problem of first story detection, which can be reformulated as the task of event detection in streaming data. The proposed approach uses a two-step process to identify first stories. First, a subset of previously observed similar tweets is identified using locally sensitive hashing. This process is executed in constant time and space. Second, a clustering approach called Threading is used to group related tweets using cosine similarity. The first tweet within a chain of tweets is presented as the first story. From a collection of 160 million tweets, the authors extract a smaller subset of tweets citing the constraints in manually preparing a Gold Standard for events. The cost of applying the second step on a large Twitter stream is not discussed and the authors do not tackle the problem of an expanding vocabulary. The approach is also restricted to ASCII formatted streams as the authors discard all other characters. Hence, it cannot be applied to multilingual streams.

Visualization of events and temporal trends of topics in text is another active area of research. Marcus et al. [39] created TwitInfo to visualize events from microblogs in real-time. Relying on user-defined events, the system generated and visualized different views of the microblog text and metadata for exploration. [30] introduced TextFlow, a system to visualize the evolution of topics in text data. They combine a Hierarchical Dirichlet Process (HDP) mixture model to model the text corpus and extracted critical events in the identified topics. Deeper exploration of the data was enabled by extracting top syntactic and semantic keywords. Wei et al. [51], built a system to visualize topical evolution in text documents. Given a user query, the system summarized document matches through topics represented by keywords. Other systems like MemeTracker [38], focus on phrases, called memes and visualize their evolution over time. Temporal evolution over documents was also visualized by ThemeRiver [33]. [31] created a system, LeadLine, to help users detect and explore events. Using microblog data from Occupy Wall Street they showed the effectiveness of the system in creating simple narratives of large text corpus. [41] showed that using their system TweetXplorer, the users can investigate events along the three facets: Who, Where, and When. Event summarization is another area which focuses on the presentation aspect of an event. In [28], the authors used HMMs to segment and summarize long-running and repetitive events such as sporting events by selecting a subset of tweets.

During a real-world event, people use Twitter to tweet their experiences. Depending on the locality and severity of the event, the number of Twitter users who tweet varies. As Twitter is a global medium, information from a few sources is retweeted by several others to disseminate the information. Typically, clustering has been employed to group related tweets together into clusters, to perform the task of event detection in both traditional and social media as discussed in

Table 1. Events Detected in the Earthquake Dataset

Day	Frequent Tweet	Support	What	Who	When (MST)	Where
Sept 5, 2011	t @quakemonitor: #earthquake m 6.5, northern sumatra, indonesia depth: 52.30 km sep 5 18:55:09 2011 bst http://t.co/mfimkje	1,519	indonesia, sumatra, depth, northern, islands		2011-09-05 01:26:21	Malaysia, Indonesia
Oct 23, 2011	rt @gma: #breaking #turkey: usgs says #earthquake has preliminary magnitude of 7.3; located in van province, on iranian border	508	#turkey, eastern, turkey, magnitude, earthquake		2011-10-22 17:25:02	
Nov 9, 2011	rt @cnnbrk: 5.7 magnitude #earthquake rocks eastern turkey, usgs says http://t.co/bin7dnhq	307	turkey, time, expect, massive, eastern		2011-11-08 17:44:40	Northwest Georgia
Feb 6, 2012	rt @bruromars: pray for the philippines! especially on visayas who were struck by an #earthquake earlier.	10,364	pray, visayas, philippines, struck, earlier		2012-02-05 20:50:38	
Apr 11, 2012	rt @huffingtonpost: a magnitude 8.7 #earthquake has been reported off the coast of indonesia. indian ocean-wide tsunami watch is in effe ...	2,890	tsunami, indonesia, magnitude, coast, indian	Mr. Mahdi	2012-04-11 01:50:05	Malaysia, Indonesia
May 20, 2012	rt @microsatira: in emilia per il #terremoto crollata una chiesa. cosa non fa dio per non pagare l'imu.	2,735	emilia, chiesa, cosa, #ferrara, stato		2012-05-20 00:45:48	

the previous section. The challenge in detecting events in streaming data is the need to process the data continuously. Thus, we cannot cluster the tweets in multiple passes as in traditional clustering algorithms.

We have collected large and distinct datasets from Twitter over an extended period, which cover different types of events. We intentionally chose datasets from different topic domains to show the effectiveness of our approach in handling events with different volume and velocity characteristics. Twitter API provides access to approximately 1% of public tweets published in a day. Hence, we cannot access all the tweets related to an event. To ensure that our dataset contain tweets specific to the event, we focus on earthquakes due to the existing research which demonstrates the use of Twitter during earthquakes [46],[40]. Some of the events detected in Earthquake data are listed in Table 1.

3   Location Inference

We present a framework to geo tag the tweets based on the content. Our framework consists of multiple predictors based on NLP, Language Models and External Sources. Based on the quality of predictions from each predictor, a confidence or reliability score is assigned to each predictor. These scores are used while combining the predictions to generate a single meta level location for a tweet. The key premise behind our approach is that combination of predictors will perform better (accuracy as well to tag more tweets) than a single predictor. Moreover, due to reliability computations, the error prone predictors will be given less importance and also provide cues to perform model updation. We show the results of our framework on streaming Twitter data. We show that combination of predictors result in good precision, recall and F-measure. The algorithms currently supported by our framework can be broadly divided into three groups:

NLP Models: This set includes models based on Regular Expression Matching, Named Entity Recognizers and POS Tagging.

Language Models: In these models, we learn language models for each location in two different ways. First, the language model is learnt using training data and associated word distributions. Second, we develop dynamic (online) language models. The second approach requires no training data.

External Information Aware Models: Finally, we use information outside of the tweet like RSS News Feeds, Hashtag Distribution etc to assign location. For example a tweet "I am going to gym" will only be tagged by POS based models by noticing the active verb 'going' and hence, using the default user location, whereas a tweet "Olympics!! Hope to see great events" will be tagged by using external information that will map Olympics to London. Finally, a tweet like "IPL Daredevils Rock!!" will be tagged using language models to Delhi, India (Daredevils is a cricket team with homebase in New Delhi). The key thesis of the proposed framework is that each algorithm will capture different facets of the tweet and assign location and overall, the framework will have higher accuracy.

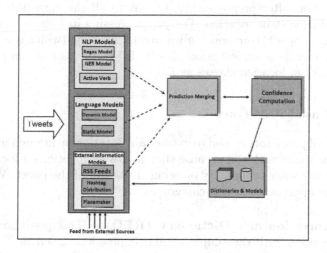

Fig. 1. High Level Overview of Our Framework

Each algorithm, in parallel, tries to identify the location for the streaming tweets. The individual predictions from each algorithm are then combined to assign location to the tweet. Our framework also incorporates *reject option* for the tweets, where a tweet is not assigned a location. This helps us to reduce false positives. Therefore, a tweet is only assigned a location when the system is confident. Another dimension of the confidence is included by dynamically assigning weights to the individual predictors based on their performance in recent past. This property allows us to bias in favor of high quality predictors and also reduces emphasis on old outdated models that tend to produce inaccurate results.

Figure 1 presents the overview of our proposed framework, which is developed on plug and play philosophy. The overall system can easily be extended to add more predictors or update the existing ones by replacing them. Moreover, with the *suite of algorithms* approach, we are able to handle lot of real world challenges for location tagging. For example, if an organization wants to periodically perform this analysis in batch mode, then the NLP and Language based models can be used while stopping the external models (as these models require real time information and news collection). Similarly, for real life evolving events (earthquake, sports, fire etc), the language models can be turned off. Finally, since all predictors are not equally fast, we perform an online continuous merging of individual predictions, and if the system can assign location with high confidence, then the other predictors can be skipped.

4   Classifier Description

In this section, we discuss the classifiers that have been used in our framework. Each tweet, after preprocessing, is sent to all the classifiers and scores are merged to predict its location. The pre-processing includes removal of stop words[1], slangs[2], top 400 common english words[3], hashtag (which are stored separately), user mentions, punctuations and links. In the rest of the tutorial, we represent the set of location classes as L.

4.1   NLP Based Predictors

This category of predictors is used to extract any location mentioned in the tweet. This predictor works with the premise that if a tweet mentions a location, then that location has some probability of being the origin of the tweet. We adopted three different approaches in this category.

Regular Expression and Dictionary (RED) - This predictor first tries to match the tweet with the simple regular expressions given in Table 2. The first entry in the table is the format of tweets generated using Foursquare (https://foursquare.com/), a popular location service applications. The second entry is a general way of mentioning a location, using the @ (alias "at") symbol. An initial dictionary of places is taken from GeoNames[4]. Whenever a new location is found using OLM and PM predictors, it is added to this dictionary. The places in the dictionary are searched in each tweet. As of now, we employ simple regular expressions for obtaining exact and partial matches. A tweet is classified into $l \in L$ if the found location is within a certain distance from l.

---

[1] http://www.textfixer.com/resources/common-english-words.txt
[2] http://www.noslang.com/dictionary/full/
[3] http://www.paulnoll.com/Books/Clear-English/English-3000-common-words.html
[4] http://www.geonames.org/

Table 2. Example Regular expressions

I'm at [Location] ([Address]) [Comment] (Link)
[Text] @ [Location] [Text]

For example, actual tweets from users like "My brother wedding (@ The Oberoi, Bangalore)" and "Good hospitality from #soekarnohatta airport staff. Sis in law @ Soekarno-Hatta International Airport (CGK)" are correctly classified into Bangalore and Jakarta, respectively.

OpenNLP Location Model (OLM) - This predictor uses the standard OpenNLP Location Model[5], trained for English sentences, to get the mentioned locations in the tweet. The models are pre-trained on various freely available corpora. The model finds locations by tokenizing the tweet into sentence and tokens. After finding a location, the same process as used for RED is used to classify the tweet into one of the classes. For example, user's tweets like, "Feels like a 10-year-old at Disneyland!!!", were correctly tagged to Los Angeles using OLM.

Active Verb Classifier (AVC) - This classifier is used only when all other classifiers fail to predict any location for a tweet. In this case, if the tweet contains an active verb, then the profile location of the user is assigned as the location. The premise behind this predictor is that tweets having active verbs are more likely to talk about the current situation of the user. Standard POS taggers are used to mark active verbs in tweets. For example, the tweet 'looove working frm home :-)' was correctly classified to Chicago, which was the profile location of the user.

4.2 Language Model Based Extractors

This section describes two approaches to model the language of the locations, which exploits the local popularity of certain words to categorize the tweets.

Fixed Language Model (FLM) - In this predictor, we develop a language model by training on tweets from different regions. We use a modeling approach similar to that in [44] to develop the location model. So, the probability of a tweet being generated from a location is estimated by sampling from the word distribution of the location. This gives the probability of each location as the origin of the tweet. The premise behind this predictor is that words like 'Statue of Liberty' would be more popular in New York than in other locations. Similarly, users tend to transliterate, i.e. use English script to tweet words from their native language. We used RED, OLM, AVC, FLM, RSS and PM classifiers for prediction. As mentioned earlier, 70% of the data was used to train FLM, and

---

[5] http://opennlp.apache.org/

rest 30% was used for experimentation. Table 3 contains the precision, recall and F-measure for the predictors.

We also studied the effect of removing one classifier at a time from the set of classifiers and using this subset for prediction, which is shown in Table 4. This result indicates the percentage of tweets that were classified only by one classifier when all others were unable to categorize the tweet or were prevented from misclassification with the help of this classifier. AVC was not removed. The results further substantiate our contention that each classifier would analyze different aspects of tweets for classification and in total would produce better results. The individual precision for each classifier is high but the reject rate is also high (as seen from low recalls). However, the recall increases to 42% when all classifiers are combined. The reduction in precision is due to poor performance of some classifiers and low applicability of good classifiers (higher rejection rates). These can be improved by incorporating better models and more exhaustive dictionaries.

Table 3. Precision, Recall and F-measure values for STATIC DATASET

Classifier	Precision	Recall	F-measure
RED	0.9891	0.0671	0.1256
OLM	0.9899	0.0669	0.1253
AVC	0.9959	0.1413	0.2474
FLM	0.8705	0.1911	0.3134
RSS	0.2216	0.0271	0.0483
PM	0.9827	0.0567	0.1072
All	**0.7628**	**0.4237**	**0.5448**

The results indicate that FLM works better than any other classifier on this data set, showing that the local words are captured effectively using the training model. If an online training method can be developed for the classifier, then this classifier would work very efficiently even for real time tweets.

Clearly, RED, PM and OLM have a very low recall score, though their precision is quite high. This is because the location model used for classification in this classifier is trained on properly formed English sentences. Tweets, however, do not have any particular format, nor are they bound to follow the syntax rules. The informality of microblogs hinders the use of such models. The high precision, however, supports our hypothesis that the mentioned location in a tweet is the location of origin of the tweet.

The precision and recall values for RSS are small as compared to the other classifiers. The reason for this observation is that very few tweets were related to events mentioned in news articles. This is reflected from the recall score of RSS. The reason for low precision of RSS might be attributed to the fact that the news articles used for this data set consisted of the daily news, and not the current news. This might have led to misclassification of tweets. The same effect is reflected by the increase in the performance parameters on removing RSS.

The drastic decrease in precision, recall and F-measure on removal of FLM or RED show that both these predictors individually classify many tweets that are not classified by any other classifier. Removal of other predictors, except RSS, also reduces the performance parameters

Table 4. Precision, Recall and F-measure on removal of one classifier on STATIC DATASET

Removed	Precision	Recall	F-measure
None	0.7628	0.4237	0.5448
RED	0.7270	0.3694	0.4898
OLM	0.7588	0.4196	0.5403
FLM	0.7897	0.3394	0.4747
RSS	0.8804	0.4339	0.5813
PM	0.7564	0.4158	0.5366

5 Faceted Summarization

Social media platforms like Twitter and Facebook have been successfully used in disseminating timely information about a whole family of events, from sporting events to political to disasters. We focus on Public Safety and Disaster events in this paper. The key distinguishing element between Public Safety and other events is that (if) the information gleaned about public safety events can be used by other citizens as well as disaster management authorities. For the information to be useful, it is extremely important that the key facts about the events are presented in an structured fashion. Recenty, microblog summarization has drawn a lot of interest [48,27,35,42,29,49].Most of the existing summarization techniques use tf-idf,n-gram matching [49,29] or augment summaries with the help of web articles and extract most relevant tweets only.

We use PoS Tagger along with Named Entity Recognizer and custom text extractors to derive the attributes from the social media stream. We show initial results on floods in Uttrakhand, India and Brazil. The Uttrakand data has around 200K Tweets where Brazil data 300K Tweets. We show some examples to demonstrate the usefulness of the approach. We have done a user survey to quantify our approach as well as comparison against some of the existing tools.

5.1 Background

In order to summarize a set of messages from Twitter, it is essential that the data be corresponding to a single event rather than spread across multiple events. Hence, the first step is to filter messages from Twitter. Since we are focussing on Public Safety and Disaster scenarios, we use keywords which are specific to public safety, e.g. flood, earthquake, etc. Monitoring of tweets from Twitter, related to such keywords, provides enough data to identify events in real time. The next step after filtering is to group the messages into clusters such that each cluster identifies a single event of interest. This clustering is done using the following attributes:

- Time of occurrence of event: Use of the time attribute when clustering is natural since the messages related to the same event will be very close temporally.
- Location of the event: Location is another important attribute used for clustering. Since we are interested in the event which is talked about in the message, we try to identify references to places or locations in the message body rather than taking the location of the sender.
- Content of the message: Finally, we consider the actual text of the message to group together similar messages. Messages belonging to the same cluster tend to share many of the keywords.

Any existing methods can be used to perform clustering of the tweets which takes into account the above criteria. We used the Event Detection and Clustering mechanism explained in previous sections.

5.2   Facet Extraction

Most of the existing work in the literature employ various methods like tf-idf, term expansion, n-grams etc. [53,29,48] to identify and output the most relevant tweets as the summary. These techniques usually fail either due to conciseness or exhaustiveness. Being too concise may miss critical information while giving too many details defeats the whole purpose of summarization.

Our current facet extraction algorithm works by looking at six facets namely *what, when, who, where, action* and *severity*. We use the Stanford Part-of-speech Tagger and Named Entity Recognizer coupled with custom text extractors to populate these facets. The process of extraction of each facet is explained in detail below.

- **What:** This facet represents the category of the public safety event like drugs, floods, shooting etc. This facet is extracted from the tweet text itself based on our list of categories of events which maybe useful for public safety.
- **Where:** Another important facet to be detected about the event is the location in order to take timely and accurate action. Previous work, generally, extract originating location from geo-tagged location present in the tweet metadata. However, the number of tweets which are geotagged represent a minisicule amount of 2% of the total tweets. In addition the location of the actual event may be different from the location of the sender. In order to make use of all the available messages, we used the tweet text to extract actual location of the event.

  In the first step, we clean up the tweet text by expanding the commonly used slangs in location names like *rd* for *road*, *st* for *street*, av for *avenue* etc. with their proper names. Next, we ran Stanford NER to detect the locations from the text. Because of the abuse of language and grammar used in some of the tweets, Stanford NER did not give precise results. We also used custom rule based text extractors coupled with regular expression patterns to capture locations from the text. An example of such a rule tags nouns which were followed by any of *at, in, on, from, near, visit, going to* as locations in the text. In the next step, we ran merging algorithm to remove noise in the locations. For example, by merging New York and York into single New York. This step eliminated most of the noise in the location facet.

- **Who:** This identifies the subject or the doer of the action in the particular event. The results for this facet was extracted by using both the Stanford NER coupled with POS tagger. Again, rules like adjective with or without nouns followed by verbs in the tweet were used to annotate the subjects in the text.

- **When:** For the time information, we were again taking tweet's metadata to extract this information. In addition to that, time mentioned in the tweet like today, now, Monday, 23 June et.c tagged by NER were also utilized.

- **Action:** Action represents action taken by the actors in the event. The verbs in the cluster tagged by POS tagger constitute this facet. This facet effectively summarises most of the content in the cluster as it gives a broad overview of the activities being talked about in the cluster.

- **Severity:** This facet is most essential for public safety and disaster events only as this represents the extent of damage like kind of loss of property and lives happened in the event. This facet will help the authoirities to gauge the extent of severity of the event to take suitable actions. This was tagged by custom text annotators which looked for numbers followed by words like *rescued, wounded, killed, murdered, missing, unconscious, stranded* etc. in the tweets. We used a sliding window of 2 words to capture these facets as there can be words like people or lives between these words.

Figure 2 shows an example faceted summary where the cluster contains tweets from people complaining about drugs being consumed in their neighborhood.

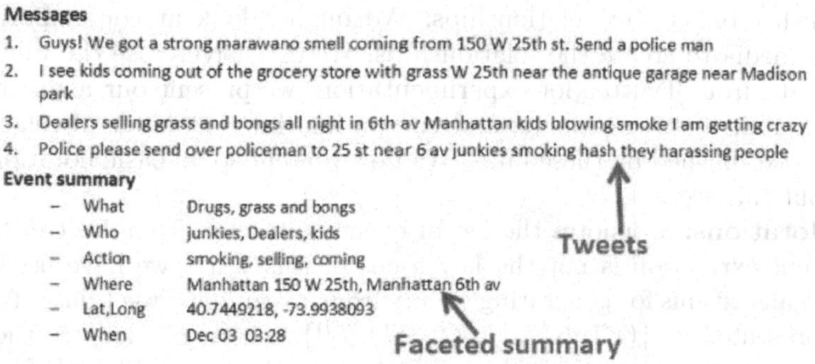

Messages

1. Guys! We got a strong marawano smell coming from 150 W 25th st. Send a police man
2. I see kids coming out of the grocery store with grass W 25th near the antique garage near Madison park
3. Dealers selling grass and bongs all night in 6th av Manhattan kids blowing smoke I am getting crazy
4. Police please send over policeman to 25 st near 6 av junkies smoking hash they harassing people

Event summary

– What	Drugs, grass and bongs
– Who	junkies, Dealers, kids
– Action	smoking, selling, coming
– Where	Manhattan 150 W 25th, Manhattan 6th av
– Lat,Long	40.7449218, -73.9938093
– When	Dec 03 03:28

Tweets

Faceted summary

Fig. 2. Example Summary

5.3   Facet Confidence

The above algorithm extracts numerous facet values per cluster depending on the size of event which can be cumbersome to read and may miss out on the importance of each facet value. To capture this, we used confidence metric which is the normalized weight of frequency of the facet value in the cluster for each facet. This determines the importance of each facet value with respect to the cluster. The confidence score was also used while presenting the facet values as these were shown in decreasing order of their confidence score.

6   Semantic Linking

In a world of millions of people (Twitter subscribers) and inherent entropy, conversations around any event are expected to move towards different directions over time. It is an extremely challenging, if not impossible, task to connect these different facets as a sequential flow of the same dialogue, or an associated dialogue, using the well-known text clustering techniques. Incorporating externally obtained domain knowledge becomes imperative to establish connections across event facet clusters. This approach of using external domain knowledge is well-known for domains with established corpus.

However, there are two significant shortcomings to this approach with respect to social microblogging networks. (1) The standard corpus-based approach does not leverage the *social knowledge* already present in the microblogging platform. (2) And, trending events often get built around non-traditional, temporary and contemporary factors, entities and relationships, so it is impractical to expect well-defined corpus documents to exist a priori. As an example, political turmoils have existed for ages; however, one may not expect a dedicated corpus to exist specifically for the Libya 2011 turmoil associating its places and locations, its contemporary leaders, all the other worldwide policitally related factors.

In this work, we attempt to overcome these shortcomings. We add knowledge from social network graph on the microblogging platform by incorporating strengths of social relationships among people tweeting around the event facets to establish cross-cluster relationships. We further look at contemporary online news media to create the relationships. We extensively use the Libya 2011 turmoil data from Twitter for experimentation. we present our algorithms for generating links between events, and comparing different kinds of relationship structures established by these links. We now present some basic notations used throughout this section.

Basic Notations: $\mathcal{E}$ denotes the list of events extracted from Twitter Stream. Since, event extraction is not the key focus of this work, we have used online clustering algorithms for generating events from streaming Tweet data. An event E^i is represented as $\{(K_1^i, K_2^i, \dots, K_n^i), [T_s^i, T_e^i]\}$, where K^i denotes the set of keywords extracted from the tweets which form the event E^i and T^i is time period of the event. We use existing established methods for computing K and T. K contains *idf* vector and proper nouns (extracted by PoS tagging) from the tweets.T is simply the time of first and last tweet in the event cluster.

Goal: The overall goal is to extract discussion sequences on microblogs, and identify the social discussion threads. This is achieved by generating an Event Graph $\mathcal{G} = \{\mathcal{E}, R\}$, where $\mathcal{E}$ represents the events and act as nodes in the graph, and then analyzing relationship edges across event clusters. The set R represents the relationship edges between events. Our algorithms use *temporal, social* and *extended semantic* relationships to identify social discussion threads from the microblogs that are otherwise unstructured and uncategorized.

6.1   Overview of Relationships

We now provide details on the extended semantic, social and temporal relationships and their extraction algorithms. To make this article self contained, we present an overview of these relationships.

Extended Semantic Relationships: This relationship is extremely useful but challenging to establish. Lets us motivate the need for such relationship by a simple example. Consider two events with associated keywords $E_1 = \{$damage, earthquake, dead, toll$\}$ and $E_2 = \{$earthquake, relief, shelter$\}$. Now, lets pick one work from each set *damage* and *relief*. One cannot establish any of the widely accepted relationships like synonymns, antonymns, hypernymns, hyposnyms etc when the words are taken in isolation. However, coupled with prior knowledge about the larger event *earthquake*, the words can be semantically related. In essence (with abuse of notation and terminology), damage and relief are independent variables without extra information, however, they are related given *earthquake*. Therefore, we would like to add the semantic edge between these events. We use external corpus to extract and quantify such semantic relationships. The corpus could be news stories, articles or books.

Social Relationships: Direct social connections are the core constituent elements of social relationships. Higher order social relationships can be established by exploring the social network structure. Well-defined structures such as communities with maximum modularity ([6], [9]) can be extracted using efficient modularity maximization algorithms such as BGLL [3].

Temporal Relationships: Allen [2] presents an exhaustive list of temporal relationships which can exist given two time periods (events in our case). The relationships include *overlap*: partof event A and event B co-occur, *meets*: event A starts as soon as event B stops, *disjoint*: event A and event B share no common time point.

6.2   Relationship Extraction

We now present the key steps of our extraction algorithm in detail. We extract extended semantic relationships, social relationships and temporal relationships.

Extended Semantic Relationship Extraction
We establish weighted extended semantic relationships across event clusters by the following steps. The input to the extended semantic relationship extraction algorithm for two events E^i and E^j is keyword list K^i and K^j.

Step 1: Generating Pairs and Pruning Mechanism - We generate $|K^i| \times |K^j|$ pairs of keywords which need to be evaluated for extended semantic relationship. To avoid similar keywords that would skew the results, we prune pairs which are related semantically (synonyms, antonyms, hypernymns and hyponynms). Since POS tagging is done on the tweets in the event, we also remove pairs where one of the words is a Proper Noun or Active Verb. We use methods such as Leacock Chodorow [12], Wu Palmer [21], Resnik [18] and Lin [13] to compute similarity, using the Wordnet lexical database. We look at the similarity scores of K^i and K^j as found by the above methods. We retain a pair of words if the similarity score S_{ij} is lesser than a desirable similarity threshold S, and prune the pair otherwise.

Step 2: Document Corpus Generation and Searching - We use the keywords used for filtering Twitter Public API to search for news stories for the same time period on contemporary external news documents. For experimentation purposes, we used Google News for retrieving relevant stories. The retrieved news stories act as our external corpus. We create an inverted index for this corpus, where for each word we store the document ids and the *tf-idf* of the word in the documents. Given the pair of words (K^i_l, K^j_m) (we will omit subscript l and m, when there is no ambiguity), we find the intersection of corresponding document lists. Therefore, at the end of this step we have list of documents (denoted by D_{lm}) in which both the words co-occur along with their *tf-idf* in the documents.

Step 3: Pairwise Score Computation - For each of the selected documents, we compute the coupling of the pair of words. Assume, $C(K^i_l, D_t)$ gives the *tf-idf* score of word K^i in document D_t. The pairwise coupling can be computed as minimum $(C(K^i_l, D_t), C(K^j_m, D_t))$. The overall coupling is calculated as average of coupling over all documents.

Step 4: Overall Score Computation - This process is repeated for all pair of words in E^i and E^j. Finally, for a given pair of event clusters E^i and E^j, if w_{ij} pairs of keywords were pruned and the rest were retained, then the overall score is $\frac{\sum_{K^i, K^j} Coupling}{(|K^i| \times |K^j| - w_{ij})}$. The final scores are ranked in descending order and top K% are selected based on user preference or can be pruned based on threshold.

Social Relationship Extraction

We construct social linkage graphs between pairs of events using social connections of event cluster members to construct edges. Each event associates a number of microblog posts (tweets) from a set of members of the microblog network (Twitter).

Given a person P and an event cluster E^i, $P \in E^i$ iff ($\exists M$), a microblog post, made by P, such that $M \in E^i$. Please note that with this definition, a person can potentially belong to multiple event clusters at the same time.

The connections are established by participation of direct social neighbors of individuals across multiple events. We draw an edge across a given pair of events if there is at least one direct (one-hop) neighbor in each event belonging to the pair of events. The weight of an edge between event cluster E^i and E^j

is determined by the total number of one-hop neighbors existing between these two clusters. So if E^i has P^i memberships, E^j has P^j memberships, the average number of neighbors in E^j of a member belonging to E^i is a_{ij} and the average number of neighbors in E^i of a member belonging to E^j is a_{ji} then the strength of the social edge between E^i and E^j is $(P^i.a_{ij} + P^j.a_{ji})$.

Temporal Relationship Extraction
The third kind of relationship we attempt to extract is temporal relationship. We look at two kinds of temporal relationships. (a) We draw a temporal edge from event E^i to event E^j if E^i ended within a threshold time gap before E^j started. For experimentation, we restrict the maximum time gap limit to a realistic 2 days. This follows from the assumption that on microblogging services like Twitter, a discussion thread will not last longer than this. This thresholding also prevents the occurrence of spurious edges across different clusters. It captures the *meets* and *disjoint* relationships described by [2]. We call this a *follows* temporal relationship. (b) We draw a temporal edge from event E^i to event E^j if E^i started before E^j started, and ended after the start but before the end of E^j. This captures the *overlaps* relationship described by [2]. Please note that unlike the undirected semantic and social relationship edges, a temporal relationship edge is always directed but unweighted. The direction is from the event with the earlier starting time to the one with the later starting time.

6.3   Social Discussion Thread Detection

After this, we follow a two-step process to identify topically evolving social discussion threads.

1. **Constructing the Semantic AND Temporal graph:** For each of the semantic and temporal graphs, we take an edge set intersection considering the directions. As the semantic graph is undirected, the direction of the temporal graph implies the direction of the resulting graph G_{ST}. Intuitively, this graph can be thought to comprise of temporally related **discussion sequence** subgraphs.
2. **Constructing the Semantic AND Temporal AND Social graph:** We perform an edge set intersection of the above graph with the social graph. This results in retaining the discussion sequences that are socially connected and eliminating the discussion sequences that are socially disconnected. The retained discussion sequences show the social evolution of discussion topics around events on microblogs. Hence, these socially connected topically evolving discussion sequences are identified as **social discussion threads**.

Measuring the Goodness of Our Approach
We attempt to quantify evolution of the topically evolving discussion sequences along the social dimension by examining the link structures. We discover BGLL communities on the social and extended semantic relationships independently; and then find the normalized mutual information (NMI: [7]) within each of these

two sets of events. Please note that NMI values range between 0 and 1, and a higher NMI value indicates a higher overlap of the two sets of communities. Our experiments indicate extremely low overlap between the communities formed around these two types of graphs.

We further discover BGLL communities on discussion sequences and social discussion threads independently, and perform NMI computation on the resulting communities to compare with the earlier NMI values. In our experiments, we observe surprisingly higher NMI values in the second case, indicating significant social propensity of evolution of discussion threads.

The presence of social discussion threads in spite of low overlap of structures formed by the baseline semantic and social graphs qualitatively indicates social localization of discussion threads.

Please note that since we have constructed 4 distinct semantic relationships and 2 distinct temporal relationships, we have $4 \times 2 = 8$ graphs constructed by the overall process. However, the experimental results suggest the presence of social discussion threads of comparable numbers and quality, irrespective of the variant of relationships used for graph construction. We show indicative results later in section. This indicates the soundness of our algorithm.

We collected Twitter data from a 2011 Libya political turmoil that had created significant impact on social media. The data was collected using (Libya OR Gaddafi) as target keywords. This implies, all the tweets used for experimentation contain at least one of the two above-mentioned keywords, or both. Table 5 shows the statistics of all the 3 datasets used in our study.

Table 5. The columns in the table show a) keywords used to search Twitter to collect the dataset, b) dates for which the data was collected, c) number of tweets collected, d) number of clusters and e) number of contemporary external news documents collected

Dataset	Keywords	Timespan	Tweets	Clusters	News Docs
Libya	Libya, Gaddafi	4 - 24 Mar'11	1011716	1344	3266
Egypt	Egypt, Protest	1 - 4 Mar'11	60948	141	1753
Olympics	Olympics, Olympic	27 Jun - 13 Aug'12	2319519	299	1186

6.4   Forming the Baseline Graphs

Since event cluster detection is not the focus of our work, we have used an available online clustering algorithm to generate the event clusters from the given tweets. We now attempt to link the clusters using external text data coming from contemporary news media to form extended semantic edges. For external contemporary news text data, we select a date range corresponding to the dates that the tweets were collected and select all the sets of news items returned by Google News (http://news.google.com) within this date range for the same set of keywords used to collect the tweets. Each of these search results had 1000 unique search result pages, all of which was used as external corpus. As described in the algorithms section, we discard edges where the Wordnet similarity between

Table 6. Wordnet Similarity Thresholds: Edges with smaller values were retained

	Wordnet Similarity Measure			
Method	Leacock Chodorow	Wu Palmer	Resnik	Lin
Value	0.98	0.52	2.0	0.4

a pair of words K^i and K^j is higher than a desirable threshold for each method. The summation of *tf-idf* values over all pairs of selected words across pairs of clusters was used as edge weights. The thresholds used in our experiments are shown on (Table 6).

We form social links based upon the one-hop social neighbors (followers) of cluster members. Weight is determined by the degree of overlap. We further establish the *follows* and *overlaps* temporal links across these clusters. The temporal links are directed unlike the social and semantic links. We thus complete the basic graph construction process.

To qualitatively inspect the goodness our approach of creating extended semantic edges, we accumulated all the three sets of event clusters formed at our Libya political, Egypt political and Olympic games datasets. We executed the same algorithm to form extended semantic edges over all these cluster sets, using all the news articles pertaining to the 3 datasets.

For visualization (Figure 3), we took random 5-6 nodes from each dataset, (17 nodes in total). The segregation among the two political datasets and the sports dataset is apparent. We find significant cross-dataset edge density across the Libya and Egypt political turmoil datasets, compared to the edge density of cross-dataset edges of Egypt or Libya with the Olympics dataset. The average edge thickness, corresponding to the extended semantic match value, is also the highest for the Libya and Egypt dataset pair. We do not show intra-cluster edges, which re denser and thicker, for ease of visualization. This qualitatively demonstrates the soundness of our algorithm to find extended semantic edges.

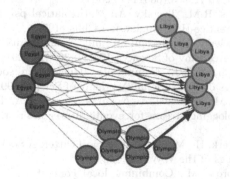

Fig. 3. Cross-dataset extended semantic edges: Visualization of a random sample

7   Conclusion

In this tutorial, we attempted to present a spectrum of research problems in Social Media Analytics. We motivated the need for analytics which can handles large volumes of fast incoming data which is written in informal language and is noisy. In particular, we focused on Event Detection (via clustering), summarization, location inferening (which is a critical task for location based services) and fusion of external information with Social Media data. We showcased Public Safety Event Detection and Linking Demonstration. In no way this is a complete survey on social media, there are other interesting research problems like how information spreads in social media, how to control rumours, expertise locator etc which are not covered here.

Acknowledgement. The authors would like to thank Shamanth Kumar and Huan Liu (ASU), Srijan Kumar, Tanveer A Faruquie, Kuntal Dey, Seema Nagar, Bhupesh Chawda, Kanika Narang and Hima Karanam for discussions and joint work which resulted in proposed algorithms and ideas.

References

1. Abrol, S., Khan, L.: Twinner: Understanding news queries with geo-content using twitter. In: Proceedings of the GIS (2010)
2. Allen, J.F.: Maintaining Knowledge about Temporal Intervals. In: Communications of the ACM (1983)
3. Blondel, V.D., Guillaume, J.L., Lambiotte, R., Lefebvre, E.: Fast unfolding of communities in large networks. J. Stat. Mech. P10008 (2008)
4. Choueka, Y.: Looking for needles in a haystack or locating interesting collocational expressions in large textual databases. In: Proceedings of the RIAO (1988)
5. Church, K.W., Hanks, P.: Word association norms, mutual information and lexicography. In: Proceedings of the ACL (1989)
6. Clauset, A., Newman, M.E.J., Moore, C.: Finding community structure in very large networks. Phys. Rev. E 70(066111) (2004)
7. Coombs, C.H., Dawes, R.M., Tversky, A.: Mathematical psychology: An elementary introduction. Prentice-Hall, Englewood Cliffs (1970)
8. Fortunato, S., Barthelemy, M.: Resolution limit in community detection. Proceedings of the National Academy of Sciences 104(1), 36–41 (2007)
9. Girvan, M., Newman, M.E.J.: Community structure in social and biological networks. Proceedings of the National Academy of Sciences 99(7821) (2002)
10. Grinev, M., Grineva, M., Boldakov, A., Novak, L., Syssoev, A., Lizorkin, D.: Tweetsieve: Sifting microblogging stream for events of user interest. In: Proceedings of the SIGIR (2009)
11. Kwak, H., Lee, C., Park, H., Moon, S.: What is Twitter, A Social Media or A News Media. In: Proceedings of the WWW (2010)
12. Leacock, C., Chodorow, M.: Combining local context and WordNet similarity for word sense identification. In: WordNet: An Electronic Lexical Database, pp. 265–283 (1998)
13. Lin, D.: An information-theoretic definition of similarity. In: Proceedings of the International Conference on Machine Learning (1998)

14. Liu, H., Singh, P.: ConceptNet - A Practical Commonsense Reasoning Tool-Kit. BT Technology Journal 22(4) (2004)
15. Nagar, S., Seth, A., Joshi, A.: Characterization of Social Media Response to Natural Disasters. In: Proceedings of the WWW (2012)
16. Pathak, N., DeLong, C., Banerjee, A., Erickson, K.: Social topics models for community extraction. In: Proceedings of the 2nd SNA-KDD Workshop (2008)
17. Porter, M.A., Onnela, J.P., Mucha, P.J.: Communities in networks. Notices of the American Mathematical Society 56(9), 1082–1097 (2009)
18. Resnik, P.: Using information content to evaluate semantic similarity in a taxonomy. In: Proceedings of the 14th International Joint Conference on Artificial Intelligence, pp. 448–453 (1995)
19. Sachan, M., Contractor, D., Faruquie, T.A., Subramaniam, L.V.: Using Content and Interactions for Discovering Communities in Social Networks. In: Proceedings of the WWW (2012)
20. Sahlgren, M., Karlgren, J.: Terminology mining in social media. In: Proceedings of the CIKM (2009)
21. Wu, Z., Palmer, M.: Verb semantics and lexical selection. 32nd Annual Meeting of the Association for Computational Linguistics, 133–138 (1994)
22. Zhou, D., Manavoglu, E., Li, J., Giles, C.L., Zha, H.: Probabilistic models for discovering e-communities. In: Proceedings of the WWW (2006)
23. Allan, J., Carbonell, J.G., Doddington, G., Yamron, J., Yang, Y.: Topic Detection and Tracking Pilot Study Final Report (1998)
24. Allan, J., Papka, R., Lavrenko, V.: On-Line New Event Detection and Tracking. In: Proceedings of the 21st Annual International ACM SIGIR Conference on Research and Development in Information Retrieval. ACM, pp. 37–45 (1998)
25. Beaumont, C.: Mumbai attacks: Twitter and Flickr Used to Break News (November 2008), http://bit.ly/1fgOgOp (accessed August 8, 2013)
26. Becker, H., Naaman, M., Gravano, L.: Learning Similarity Metrics for Event Identification in Social Media. In: Proceedings of the Third ACM International Conference on Web Search and Data Mining, WSDM, pp. 291–300. ACM (2010)
27. Beverungen, G., Kalita, J.: Evaluating methods for summarizing twitter posts. In: WSDM, pp. 11:9–11:12, (2011)
28. Chakrabarti, D., Punera, K.: Event Summarization Using Tweets. In: Fifth International AAAI Conference on Weblogs and Social Media, ICWSM (2011)
29. Chakrabarti, D., Punera, K.: Event summarization using tweets. Proceedings of the Fifth International AAAI Conference on Weblogs and Social Media, 66–73 (2011)
30. Cui, W., Liu, S., Tan, L., Shi, C., Song, Y., Gao, Z., Qu, H., Tong, X.: TextFlow: Towards Better Understanding of Evolving Topics in Text. IEEE Transactions on Visualization and Computer Graphics 17(12), 2412–2421 (2011)
31. Dou, W., Wang, X., Skau, D., Ribarsky, W., Zhou, M.X.: LeadLine: Interactive Visual Analysis of Text Data Through Event Identification and Exploration. In: 2012 IEEE Conference on Visual Analytics Science and Technology (VAST), pp. 93–102. IEEE (2012)
32. Fung, G.P.C., Yu, J.X., Yu, P.S., Lu, H.: Parameter Free Bursty Events Detection in Text Streams. In: Proceedings of the 31st International Conference on Very Large Data Bases, VLDB Endowment, pp. 181–192 (2005)
33. Havre, S., Hetzler, E., Whitney, P., Nowell, L.: Themeriver: Visualizing thematic changes in large document collections. IEEE Transactions on Visualization and Computer Graphics 8(1), 9–20 (2002)

34. Huang, C.: Facebook and Twitter key to Arab Spring uprisings: report (June 2011), http://bit.ly/io4WUG (accessed August 27, 2013)
35. Khabiri, E., Caverlee, J., Hsu, C.-F.: Summarizing user-contributed comments. In: Fifth International AAAI Conference on Weblogs and Social Media (2011)
36. Were, D.K.: How Kenya turned to social media after mall attack (2013) (accessed September 25, 2013)
37. Kumaran, G., Allan, J.: Text Classification and Named Entities for New Event Detection. In: Proceedings of the 27th Annual International ACM SIGIR Conference on Research and Development in Information Retrieval, pp. 297–304. ACM (2004)
38. Leskovec, J., Backstrom, L., Kleinberg, J.: Meme-Tracking and the Dynamics of the News Cycle. In: Proceedings of the 15th ACM SIGKDD International Conference on Knowledge Discovery and Data Mining, pp. 497–506. ACM (2009)
39. Marcus, A., Bernstein, M.S., Badar, O., Karger, D.R., Madden, S., Miller, R.C.: Twitinfo: Aggregating and Visualizing Microblogs for Event Exploration. In: Proceedings of the 2011 Annual Conference on Human Factors in Computing Systems, pp. 227–236. ACM (2011)
40. Mendoza, M., Poblete, B., Castillo, C.: Twitter Under Crisis: Can we Trust What We RT? In: Proceedings of the First Workshop on Social Media Analytics (2010)
41. Morstatter, F., Kumar, S., Liu, H., Maciejewski, R.: Understanding twitter data with tweetxplorer. In: Proceedings of the 19th ACM SIGKDD International Conference on Knowledge Discovery and Data Mining, pp. 1482–1485. ACM (2013)
42. O'Connor, B., Krieger, M., Ahn, D.: Tweetmotif: Exploratory search and topic summarization for twitter. In: Proceedings of ICWSM, pp. 2–3 (2010)
43. Petrovic, S., Osborne, M., Lavrenko, V.: Streaming First Story Detection with Application to Twitter. In: Proceedings of NAACL, vol. 10. Citeseer (2010)
44. Ponte, J.M., Croft, W.B., Croft, W.B.: A language modeling approach to information retrieval. In: SIGIR, pp. 275–281 (1998)
45. Qu, Y., Zhang, C.P., Zhang, J.: Microblogging After a Major Disaster in China: A Case Study of the 2010 Yushu Earthquake. In: CSCW, pp. 25–34 (2011)
46. Sakaki, T., Okazaki, M., Matsuo, Y.: Earthquake Shakes Twitter Users: Real-Time Event Detection by Social Sensors. In: Proceedings of the 19th International Conference on World Wide Web, WWW, pp. 851–860. ACM (2010)
47. Sayyadi, H., Hurst, M., Maykov, A.: Event Detection and Tracking in Social Streams. In: Third International AAAI Conference on Weblogs and Social Media, ICWSM (2009)
48. Sharifi, B., Hutton, M.-A., Kalita, J.K.: Experiments in microblog summarization. In: Proceedings of the 2010 IEEE Second International Conference on Social Computing, SOCIALCOM 2010, pp. 49–56 (2010)
49. Summers, E., Stephens, K.: Politwitics: Summarization of political tweets (2012)
50. Tsukayama, H.: Twitter turns 7: Users send over 400 million tweets per day (2013), http://articles.washingtonpost.com/2013-03-21/business/37889387_1_tweets-jack-dorsey-twitter (accessed September 25 2013)
51. Wei, F., Liu, S., Song, Y., Pan, S., Zhou, M.X., Qian, W., Shi, L., Tan, L., Zhang, Q.: Tiara: A Visual Exploratory Text Analytic System. In: Proceedings of the 16th ACM SIGKDD International Conference on Knowledge Discovery and Data Mining, pp. 153–162. ACM (2010)
52. Weng, J., Lee, B.S.: Event detection in twitter. In: Fifth International AAAI Conference on Weblogs and Social Media, ICWSM (2011)

53. Ghotig, A., Yang, X., et al.: A framework for summarizing and analyzing twitter feeds. In: SIGKDD (2012)
54. Yang, Y., Pierce, T., Carbonell, J.: A study of retrospective and on-line event detection. In: Proceedings of the 21st Annual International ACM SIGIR Conference on Research and Development in Information Retrieval, pp. 28–36. ACM (1998)
55. Zhang, K., Zi, J., Wu, L.G.: New Event Detection Based on Indexing-Tree and Named Entity. In: Proceedings of the 30th Annual International ACM SIGIR Conference on Research and Development in Information Retrieval, pp. 215–222. ACM (2007)

Temporal Analysis of User Behavior and Topic Evolution on Twitter

Mona Jain[1], S. Rajyalakshmi[3], Rudra M. Tripathy[2], and Amitabha Bagchi[1]

[1] Indian Institute of Technology, Delhi, India
[2] Silicon Institute of Technology, Bhubaneswar, India
[3] Unaffiliated

Abstract. We investigate the temporal aspects of user behavior in this paper and relate it to the evolution of particular topics being discussed on the Twitter OSN. Studies have shown that a small number of frequent users are responsible for the maximum percentage of tweets. We further hypothesize that users deviate from their usual tweeting hours when a major event occurs. With these as our underlying concepts, we introduce a new metric called "tweet strength" that gives more weight to tweets by users who in general tweet less or those who do not usually tweet at a given time. We study the evolution of a set of topics through the lens of tweet strength and try to identify the classes of users driving the popularity at different times. We also study word-of-mouth diffusion mechanism through the network by defining a "copying" behavior. When a follower of a user tweets on the same topic as the user, the follower is said to have copied. We further make the definition time-dependent by imposing temporal thresholds on it.

1 Intoduction

The most popular form of temporal analysis of a topic on online social networks has been to study how the number of users speaking on the topic evolves with time. Though this is an immediate evaluation and relates directly to the popularity of the topic, it does not give us any more information about the users who spoke on the topic. For instance, one may like to know, at least in hindsight, about the kinds of groups which contributed to virality of the topic. Was the news media the key to driving the topic viral? Or did the fans/ students of the topic make it popular? Even for non-viral topics, one might like to know, what are the kinds of users who normally tweet on the topic. This would help in identifying the target users, given a purpose.

We propose here, a novel method of achieving such insight without explicitly analysing the interests of every user individually. We come up with the concept of "tweet strength" which quantifies tweets not just in terms of number but also in terms of behavior of users who tweeted them. We define tweet strength as the inverse of the number of tweets by a user on an average in a given time slot. The idea behind this formulation is to classify regular users who tweet massively and infrequent users who do not tweet that often. Frequent users tend to tweet

V. Bhatnagar and S. Srinivasa (Eds.): BDA 2013, LNCS 8302, pp. 22–36, 2013.
© Springer International Publishing Switzerland 2013

on many topics and hence are not an ideal representation of users interested in the topic. Such massive users account for a large percentage of tweets at any given time. However, they are a small minority of the entire user network. The major chunk of it consists of infrequent users who may follow many frequent users, but themselves choose to tweet selectively. It is this class of users which magnifies a news by picking it up from the frequent users, thereby declaring a definite interest of the audience in the news. Even though the infrequent users may have a small following, the declaration of interest by an infrequent user is copied rapidly leading to a buzz in his local community of users. The class of infrequent users also contains a few celebrities who have a huge following but tweet very little.

Thus, frequent users can be viewed as broadcasters and news media equivalents in the Twitter network while the infrequent users act as the audience. Giving a greater tweet strength to the infrequent users amplifies the effect of a few tweets by them as against many tweets by a frequent user on the topic. Further, we consider the tweet strength of a user not as an absolute measure but as a metric dependent on the time he tweets. We divide the day into time buckets and define tweet strength of a user based on the number of times he tweets in each of these buckets. Thus, even for a frequent user, tweeting at an odd time, when he normally doesn't tweet, fetches his tweet a greater strength. The methodology of bucketing allows us to capture anomaly when an event occurs. This would be of specific relevance in emergency situations such as natural disasters, attacks, epidemics etc.

The integral role of infrequent users in the popularity of a topic leads us to investigate how the infrequent users pick topics. The only way of exposure on the Twitter network is by reading the tweets of users whom one follows. The dynamic nature of the Twitter "timeline" allows a user only so much time to note a tweet. As new tweets pour into the timeline, the user misses out on the older ones as there are only as much tweets as he can scroll through. When he finds a tweet interesting, he almost always tweets on it immediately. The decision of the user could be influenced by myriad factors like who tweeted it, the strength of the tie with that neighbour etc. While we will not go into why the user chooses to tweet on it, we take with us the fact that the user "copies" the tweet almost immediately before he loses the tweet. We consider here, the copying behavior in as fundamental a form as discussed above. Instead of restricting ourselves to Twitter specific features like RTs, MTs etc., we only focus on the topic of the tweet and whether the user picked it up from one of the users he follows.

Thus, we work on the following guidelines.

- **Assumption 1:** *The media handles tweet regularly and massively while the general audience tweets less frequently.*
  We exploit this assumption to use frequency of the user as a quick estimate of the class a tweet belongs to.
- **Assumption 2:** *The audience which tweets less frequently than the media is responsible for driving virality.*

Greater the number of infrequent users tweeting on the topic, greater are the chances of the topic going viral.

– **Assumption 3:** *Any deviation from the usual tweeting behavior of an individual user is an indication of a buzz in his community.*
The aggregate user behavior has been observed to follow some regular temporal patterns. [1]. This suggests that some users may also individually follow a tweeting pattern of their own. A tweet at an unusual hour for the user is indicative of the immediate importance of the topic for the user.

The two main Sections in the paper wish to answer the following questions:

– How can the temporal user behavior provide more information about the true popularity of a topic in the network?
– Is a short time window before a tweet enough to characterise the copy behavior of the user? If yes, then is it also a faithful representation of the aggregate activity on a particular topic?

2   Related Work

Evolution of topics on the blogosphere has been extensively investigated. Gruhl and Guha [2] propose a topic on the blogspace to consist of chatter and spikes. They argue that spikes are caused by exogenous events. When an endogenous event triggers a sharp reaction from the bloggers, the chatter sees a spike which they term as resonance. They thus classify topics as - just spike, spiky chatter and mostly chatter. They break down topic evolution into different regimes. To identify key users in these regimes, they use a random user as a baseline. We however, define surprising activity by a user based on his own behavior as opposed to comparing him with a random user.

Yang and Leskovec [3] explore the temporal patterns of popularity of online content. Kleinberg et al [4] study evolution of topics on the blogosphere. They consider two main ingredients in modeling dynamics of news cycle - imitation and recency. Our definition of copy takes both into account. Further, instead of the users opting for recent topics, the Twitter timeline forces them to copy only from the recent past. We consider here recency with respect to the tweets the user is exposed to and not in terms of the time the topic came into the network or the time the user first heard about it.

Virality in blogosphere has also been studied by Nahon et al [5]. They classify blogs into four categories - elite, top political, top general and tail blogs. They show that elite and top general blogs generate viral information and drive virality while the top political and tail blogs are followers who sustain the virality. Crane et al [6] study virality of videos on Youtube. They categorize videos as viral, quality and junk. The viral videos spread through word-of-mouth. Quality videos are similar to viral but experience a sharper rise in activity owing to their inherent quality. Junk videos experience a burst for some reason but do not spread through the network.

The microblogging networks have also been studied for user behavior and virality. Extensive mesurement study has been conducted by Sysomos [1]. Gummadi et al [7] study distribution of mentions and retweets across users and their temporal evolution. Galuba et al [8] study tweets mentioning URLs. They found the frequency distribution of users tweeting URLs to be a power law with an exponent of -1.25.

Several studies on Twitter specific features like @ mentions and retweets have been done. Sousa et al[9] study the @ mentions on Twitter nework for three topics - sports, religion and politics. They show that except for a few users with large ego-centric networks, the interactions are mostly based on the social aspect and not motivated by the topic. This is one of the manifestations of our assumptions that users pick up topics mostly by virtue of copying it from the people they follow. @ mentions have also been studied by Yang et al [10]

Several studies have investigated retweets as a mechanism of diffusion of information through the Twitter network. Kwak et al [11] study the topological properties of retweet trees. Since our data set is built on the data set they used (See 3 for details), our work can easily be compared. Wang et al [12] show that the rate of tweets alone is not enough to determine the popularity of a topic and argue the main reason behind popularity is retweet. Further, they found that users such as news sources, who tweet a lot, got retweeted many times but many of these either didnt make trends or didnt last long as trends. The audience thus picks up a few topics and amplifies them to make them viral. Welch et al [13] show that retweet is a better indicator of a user's interest in the topic than the users he follows. Rodrigues et al [14] found that word-of-mouth trees on Twitter were much wider than deep. They attribute this to the fact that a user does not choose his followers but instead his tweet gets broadcasted to all of them. They also report that the mean time to pass information across a link is very short and that 53% of RTs occurred within a day.

Kleinberg and Watts [15] define a temporal notion of distance as the time required for information to spread from one node to another. They find the network backbone to be a sparse subgraph consisting of edges along which information traverses fastest. Lerman and Ghosh [16] also study diffusion of information through Twitter and Digg networks. Kleinberg et al [17] show that users should be exposed multiple times to a topic at short intervals to make them adopt it. Rodrigues et al [18] observe various OSNs and study effect of usage frequency of users on attributes of the sessions such as its duration.

The activity of users has also been studied elsewhere. Cheong and Lee [19] use a machine learning based clustering tool to detect hidden patterns in attributes of users tweeting on a particular topic such as geography, age, gender, number of followers etc. Abel et al [20] develop a formulation to model the interests of users on Twitter with the objective of recommending web sites to them.

Table 1. Data set summary

Tweets	196,985,580
Users	9,801,062
Hashtags	1,341,733
Retweets	15,126,588
Direct (@) Tweets	41,951,786

3   Methodology

3.1   Dataset Description

The data set used in the measurement study in this paper is a part of the data we have engineered for an ongoing measurement project being carried out by our group. We describe the data set and the methodology only in outline here. The engineering aspects of preparing such a data set are detailed in our earlier work [21].

The data set we engineered is a part of the 'tweet7' data set which was crawled by Yang et al. [3]. We use the first three month's (11 June, 2009 to 31 August, 2009) tweets from this data set. The summary of the data set is shown in Table 1. The 'tweet7' data set contains information only about the tweets by each user and not the social relationships between the users. Therefore, to build the social graph of these users, we merged 'tweet7' data set with another data set crawled by Kwak et al.[11]. They crawled the information on the followers of almost all Twitter users during the same period as that of 'tweet7' data.

Twitter allows their users to tag the tweets with a topic by using hashtag. A hashtag is a single word preceded by a hash (#) symbol. We, however, found only 10% of our tweets tagged by hashtags. Therefore, we used OpenCalais [22], a text analysis engine, to identify the topics from the remaining 90% of tweets. Using hashtags and OpenCalais, we are able to extract nearly 6.2M topics in 52M tweets.

3.2   Timezone Shifts

The activity of a user varies with time of the day, the night being the time of lowest activity. Using geospatial data of the users, we convert the UTC time to the local time of the user.

Geographical information of each Twitter User was extracted using ShowUser developer API using Twitter4J java library for each of theTwitter users using only one whitelisted account which could execute upto 20,000 queries per hour to Twitter. It was found that 39% of users have not filled in their geographical location on Twitter. The geographic data of 61% of the users was converted into geographical coordinate system using Yahoo Placefinder API. The details of the conversion technique can be found in our previous work [21]

Using this geographical information of each user, the time offset of their respective location was added or subtracted from the UTC time provided by Twitter to give the local tweeting time of each user. Here, we have not considered the timezone of the user from which the tweet was posted but rather taken into account the timezone according to the geographical location given by the users. This might cause a bias in the results but we feel that it is true only for a very small fraction of the tweets.

3.3   Tweet Strength

We divide the 24 hours of the day into 7 time slots : 1-6, 7-9, 10-12, 13-15, 16-18, 19-21, 22-24 . Scrolling through the activity of each user, we group the tweets based on the slot they lie in according to the local time of the user. Doing thus, we get the number of times the user has tweeted in each slot during the entire time of observation. We note the day the user tweets for the first time since our observation. Subtracting this from the entire window of 80 days, we get the number of days we have effectively observed the user for. We then divide the number of tweets of the user in each slot by the number of effective days we have observed him for. Thus, we get an average number of tweets per day in each slot for a particular user.

Using this treatment of the daily behavior of a user, we go on to define "tweet srength" of a tweet by the user.

Definition (Tweet Strength): The strength of a tweet by a user at a particular timestamp is defined as the inverse of the average number of tweets by that user in a day in the time slot that the timestamp falls in.

Thus, a tweet by a user who frequently tweets in that time slot is given less strength as compared to one by an infrequent user who hardly tweets in that time slot. Our definition of tweet strength encapsulates two basic features : frequency and anomaly.

The frequency aspect: A tweet on a topic by an infrequent user is given more strength owing to the interest and weighted decision taken by him. A tweet by a frequent user who tweets on several topics is given less weightage. Categorizing users in this manner allows us to identify news media user accounts and other such broadcasters who tweet frequently and act as updaters on the social networking site. The audience on the other hand tweets less frequently and responds to the information it is exposed to. Though at any given point of time, the frequent users constitute a major percentage of tweets, while studying the evolution of a particular topic, the interest of the frequent and infrequent users cannot be predicted offhand. Further, our formulation of tweet strength allows us to identify the presence of even a few infrequent users owing to the high strength of their tweets.

The anomaly aspect: The formulation of bucketing the tweets into time slots allows us to identify any deviation from the usual behavior of a user. A tweet which lies in the time slot in which a user usually tweets, is attributed with a low

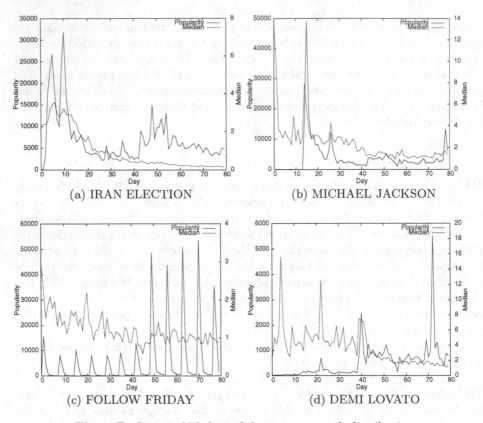

Fig. 1. Evolution of Median of the tweet strength distribution

strength. However, when the user tweets at a time when he normally doesn't, it is an indication of importance of the tweet for the user and an immediate interest of the user in the topic. This would be the scenario for the spread of a high profile rumor on the virtual network or an event in the real world, specifically emergency situations such as disasters. Such a tweet is given a greater strength value owing to its immediate significance.

Incorporating both these aspects into the strength of a tweet, we analyse the evolution of a topic based on the cumulative strength of all tweets on the topic at any given time.

4   Topic Evolution and User Behavior

We look at the evolution of a topic by observing the number of tweets on the topic with time. We define this as the popularity of the topic. With every tweet is associated a tweet strength as per the user who has tweeted on it and the time slot the tweet falls in. Thus, at any given time, we have a list of tweets on the topic and a strength associated with each of them. We study the evolution of

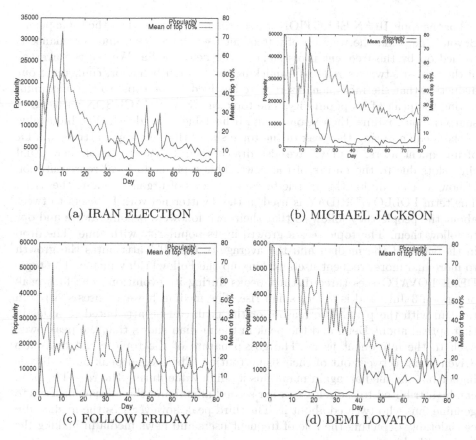

(a) IRAN ELECTION   (b) MICHAEL JACKSON

(c) FOLLOW FRIDAY   (d) DEMI LOVATO

Fig. 2. Evolution of the mean of top 10% of tweet strengths

the median of the distribution of strengths at any given instant. The median is representative of the proportion of frequent and infrequent users tweeting on the topic. A low value for the median shows that the topic is dominated by frequent users such as the news media. A rise in its value indicates that the topic has caught the attention of the general audience of infrequent users who now account for a higher proportion of tweets on the topic at a given time.

4.1   Evolution of Tweet Strength

We study the statistical properties of the distribution of tweet strengths at a given time as the topic evolves. In Fig 1 we plot the median of the distribution which hints at the fraction of infrequent and frequent users tweeting on the topic. We also plot the average of the top 10% of tweet strengths of all the given tweets at that time in Fig 2. This value gives us an idea of the extent of infrequent nature of the users. For most topics, we see a constant proportion of tweets belonging to frequent and infrequent users. However, any unusual activity in a topic is captured by the metric as we show in a few select examples below.

For the topic IRAN ELECTION, we see the median rise when the topic peaks. Beyond the peak, the fraction of infrequent users goes down and sustenance is carried on by the frequent users such as the news media. We see the average of the top 10% tweet strengths mimicking the median behavior, confirming our deduction that the infrequent users were involved in the topic to the maximum during the peak of its popularity. The topic MICHAEL JACKSON witnesses a sharp rise on 25 June, the reason being Michael Jackson's death. The high value of the median during the peak of the topic shows that the topic attracted a lot of infrequent users. It may have also drawn users to tweet out of their usual time slots due to the nature of the news. This topic is an ideal representation of how we can employ this metric to capture news of urgency among the users. The term FOLLOWFRIDAY is used in the Twitter network by users to tweet about the people they follow so that their own followers become aware and opt to follow them. The topic sees a growth in its popularity with time. The drop in the value of the median and the average of top 10% attributes the growth to more and more frequent users picking up the FollowFriday meme. The topic DEMI LOVATO sees three distinct peaks during its evolution. The first peak occurs on 3 July and is a very small rise. The median however rises sharply in tandem with the peak. The reason for the peak can be attributed to a gossip that spread about Demi and the peak in the median shows that the peak owed itself to the infrequent users. The gossip nature of the topic could have also driven users to tweet out of their usual routine. The second peak on 21 July is higher but the median again maintains its high value during the rise. The peak can be attributed to the release of her second album and the infrequent users to genuine fans who tweeted about it. The third peak however sees the median dip considerably depicting the role of frequent users and news media in making the topic peak this high.

4.2   Role of Frequent Users

While the tweet strengths highlight the involvement of infrequent users, the frequent users who keep the network alive have crucial contributions to the popularity of topics. The frequent users play a dual role in the evolution of a topic.

The frequent users constitute a major proportion of tweets at any time. While looking for topic popularity in a small time window, these users would be quintessential as they are the active participants of the network. Competition between topics prevailing at the same time would be highly influenced by these users in terms of number of tweets in the small time window. Our formulation of tweet strength will help us identify topics which are viral because of a media account promoting the topic by tweeting on it frequently. These topics need not necessarily be caught on by the general audience.

These users also tend to have more number of followers and hence act as spreaders in the network. A single tweet of theirs on the topic exposes a large number of users to the topic thereby encouraging them to talk on it too. These users also tend to be the connectors between different groups of homophilic users.

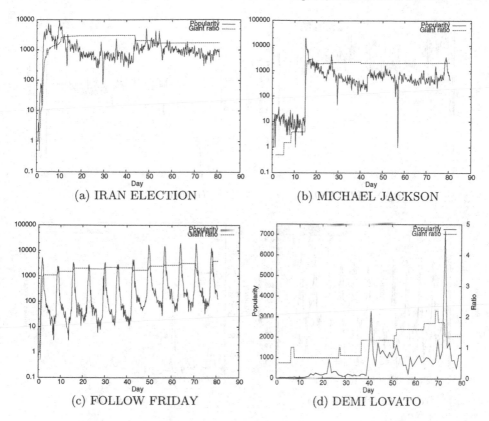

(a) IRAN ELECTION

(b) MICHAEL JACKSON

(c) FOLLOW FRIDAY

(d) DEMI LOVATO

Fig. 3. Evolution of the ratio of the largest strongly connected component to the second largest

Owing to their large follower count, they do not belong to a close knit group sharing a specific interest but instead span a wider interest range.

This leads us to the basic question of how information spreads on the Twitter network. When a tweet in the timeline of a user encourages him enough to himself tweet on the topic, the topic can be said to have spread. In other words, when users copy topics from tweets, the topic spreads. Copying the topic, synonymous with word-of-mouth, helps spread a news or rumor within the network. In the next section, we define a copy and look at its relevance in the evolution of a topic.

5   Diffusion of a Topic

The spread of a topic is essential to evaluate its true popularity. While certain topics like natural disasters or panic events can make numerous people independently tweet on the topic simultaneously, for most of the other topics, the spread is by word of mouth. Many studies on RTs, @ mentions etc. have been

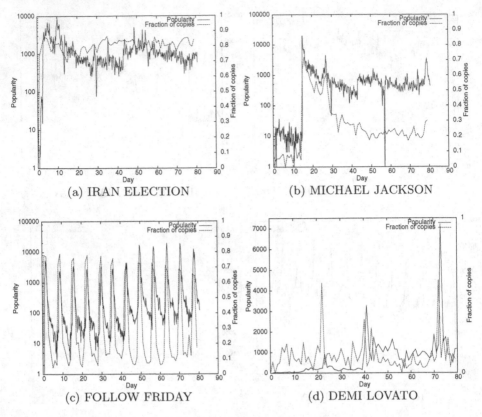

Fig. 4. Evolution of fraction of copy tweets

done in an attempt to observe how topics spread from one user to another. RTs, MTs etc hold a big challenge though. The change in the words of the original tweet, the lack of acknowledgement by a user retweeting the tweet and the self added thoughts of the user on the topic change the nature of the original tweet and make it difficult to track copies down. We consider here a more general approach towards copy, which does not look at copies of tweets but instead focuses on copies of a topic.

The notion behind copy is fundamental in nature. When a user sees a topic in his timeline, if he too chooses to tweet on the topic, we consider it a copy. Even though his tweet may not be a retweet, the original tweet encouraged him to speak on the topic and hence is a spread of the topic.

A user can only scroll through or sit through a certain number of tweets. The timeline on the Twitter interface refers to the set of tweets for a user by all the users he follows (his followees). For a user to copy a topic from a tweet in his timeline, the tweet should be within this threshold. Thus, we consider only those original tweets which lie within this past threshold of tweets in the user's timeline.

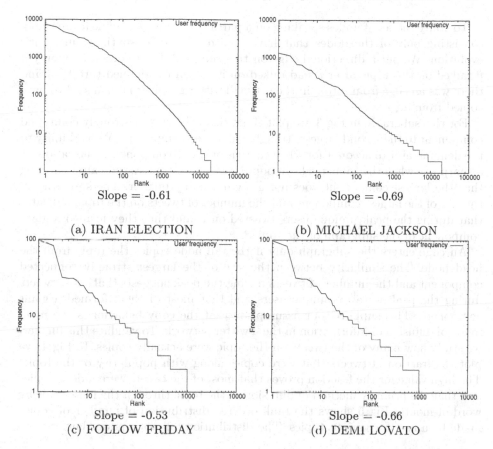

Slope = -0.61

(a) IRAN ELECTION

Slope = -0.69

(b) MICHAEL JACKSON

Slope = -0.53

(c) FOLLOW FRIDAY

Slope = -0.66

(d) DEMI LOVATO

Fig. 5. Rank ordered distribution of number of tweets on the topic per user

Definition of Copy: We declare a tweet as a copy if a user speaks on a topic within a threshold T number of tweets of all his followees put together.

With this definition of copy in place, the dual role of frequent users becomes evident. Owing to their frequent tweeting, they have a greater tendency to copy topics from their timeline and hence contribute actively to the spread of the topic. On the other hand, owing to their larger number of followers, they expose more users to the topic thereby encouraging them to copy the topic. Owing to their frequency, their tweets are likely to lie within the threshold of past tweets for their followers. If these tweets happen to be on the topic of interest, the probability of copying the topics goes up even more. In the graphs that follow, we take the threshold T to be 50 tweets. [1]

---

[1] We considered several thresholds and found that the acts of copy saturated beyond 50 and hence it was a suitable choice.

To study the set of nodes speaking on a particular topic, we define a subgraph, consisting only of the nodes that have copied this topic anytime during it's evolution. We put a directional edge on this subgraph from node u to v only if v talked on this topic after u had talked on it within the defined threshold and there was an edge from u to v in the actual Twitter graph. v is then said to have copied from u.

For this subgraph, in Fig 3, we plot the ratio of the largest strongly connected component to the second largest strongly connected component. We add unity to the denominator to account for when there is only one component. The ratio sees a sharp rise when the topic peaks in popularity and saturates thereafter showing that the largest component does not grow much any further. The similarity in the size of the largest component and the number of tweets on the topic indicates that during the peak, unique users tweeted on it and that they formed a giant component in the subgraph of the topic.

An edge enters the subgraph only if the tail node copies the topic from the head node. The similarity between the size of the largest strongly connected component and the number of tweets during the peak suggests that most tweets during the peak were by unique users and that most of them formed a giant component. This supports our assumption that the copy behavior is the main mode of diffusion of information in the Twitter network. To validate this further, we study how many of the tweets on the topic were actually copies. In Fig 4, we plot the fraction of tweets that were copies along with popularity of the topic. The high value for the fraction proves that most of the tweets were indeed copies and that the primary method of diffusion of the topic through the network is by word-of-mouth. Fig 5 shows the rank ordered distribution of number of copies made by a user for various topics. The distribution follows a power law.

6   Conclusions

In this paper, we have studied the user behavior pattern on Twitter and based on their frequency of usage of the OSN, we assigned a strength to their tweet. A tweet by a more frequent user came with less strength as compared to an infrequent one. Our formulation of bucketing the tweets into time slots defines infrequency of a user with respect to a time slot. Doing so, we were able to identify the user class behind the evolution peaks in the popularity of topics. Such a treatment can be effectively employed to identify anomaly events such as disasters etc.

We also study diffusion through the Twitter network by a fundamental approach towards copying behavior. Instead of retweets, we look at the topic of the tweet. By imposing threshold on the number of tweets a user gets exposed to before he copies the topic from an earlier tweet, we were able to capture the recency effect in the word of mouth mechanism of diffusion.

Having validated the short time window responsible for the decision of a user to copy a topic, the question which arises is - is it a hindrance to the topic diffusion? The answer we come up with is no. The short time window is a practical manifestation of the limited attention of the users. It is this time window that leads to active competition between the existent topics at a given time. Further, since the topics need to catch the attention of the infrequent users, they not only need to reach these users but also need to persist in their timelines. This suggests that a topic cannot afford to saturate a local community of users and then spread out. It instead should place seeds in several communities which then coalesce as infrequent users in each community copy the topic.

The short time window of the copy behavior also brings with it a glimpse of a limit on the number of followees a user can have. Following a large number of users would mean missing out on many tweets. A user thus follows only as much people as his own tweeting frequency permits him to. Further, the users he follows are inconspicuously chosen according to their tweeting behavior. Only a small fraction of frequent users are followed to avoid floodig of the timeline. The activity patterns of individual users is also essential for communication between two infrequent users. Their infrequency of usage leads to a time gap in the transfer of information, whic h may then take a route through their common neighours. This time lag is crucial for deciding truly trending topics. Choosing a small time window for estimation will result in domination by frequent users thereby overlooking the true interest of the topic among the audience.

References

1. Cheng, A., Evans, M., Singh, H.: Inside Twitter: An In-Depth Look Inside the Twitter World. Technical report, Sysomos Inc. (2009)
2. Gruhl, D., Guha, R., Liben-Nowell, D., Tomkins, A.: Information diffusion through blogspace. In: Proceedings of the 13th International Conference on World Wide Web, WWW 2004, pp. 491–501. ACM, New York (2004)
3. Yang, J., Leskovec, J.: Patterns of temporal variation in online media. In: Proceedings of the Fourth ACM International Conference on Web Search and Data Mining, WSDM 2011, pp. 177–186. ACM, New York (2011)
4. Leskovec, J., Backstrom, L., Kleinberg, J.: Meme-tracking and the dynamics of the news cycle. In: Proceedings of the 15th ACM SIGKDD International Conference on Knowledge Discovery and Data Mining, KDD 2009, pp. 497–506. ACM (2009)
5. Nahon, K., Hemsley, J., Walker, S., Hussain, M.: Blogs: spinning a web of virality. In: Proceedings of the 2011 iConference, iConference 2011, pp. 348–355. ACM, Seattle (2011)
6. Crane, R., Sornette, D.: Robust dynamic classes revealed by measuring the response function of a social system. PNAS 105(41), 15649–15653 (2008)
7. Cha, M., Haddadi, H., Benevenuto, F., Gummadi, K.P.: Measuring User Influence in Twitter: The Million Follower Fallacy. In: Proceedings of the 4th International AAAI Conference on Weblogs and Social Media, ICWSM 2010 (2010)
8. Galuba, W., Aberer, K., Chakraborty, D., Despotovic, Z., Kellerer, W.: Outtweeting the Twitterers - predicting information cascades in microblogs. In: Proceedings of the 3rd Conference on Onsline Social Networks, WOSN 2010 (2010)
9. Sousa, D., Sarmento, L., Mendes Rodrigues, E.: Characterization of the Twitter @replies network: are user ties social or topical? In: Proceedings of the 2nd International Workshop on Search and Mining User-Generated Contents, SMUC 2010, pp. 63–70. ACM, Toronto (2010)

10. Yang, J., Counts, S.: Predicting the Speed, Scale, and Range of Information Diffusion in Twitter. In: 4th International AAAI Conference on Weblogs and Social Media, ICWSM (2010)
11. Kwak, H., Lee, C., Park, H., Moon, S.: What is Twitter, a social network or a news media? In: Proceedings of the 19th International Conference on World Wide Web, WWW 2010, pp. 591–600. ACM, New York (2010)
12. Asur, S., Huberman, B.A., Szabo, G., Wang, C.: Trends in Social Media: Persistence and Decay. Proceedings of the 5th International AAAI Conference on Weblogs and Social Media, ICWSM 2011 (2011)
13. Welch, M.J., Schonfeld, U., He, D., Cho, J.: Topical semantics of Twitter links. In: Proceedings of the Fourth ACM International Conference on Web Search and Data Mining (2011)
14. Rodrigues, T., Benvenuto, F., Cha, M., Gummadi, K.P., Almeida, V.: On word-of-mouth based discovery of the web. In: Proceedings of the 2011 Internet Measurement Conference, IMC 2011 (2011)
15. Kossinets, G., Kleinberg, J., Watts, D.: The structure of information pathways in a social communication network. In: Proceedings of the 14th ACM SIGKDD International Conference on Knowledge Discovery and Data Mining (2008)
16. Lerman, K., Ghosh, R.: Information Contagion: An Empirical Study of the Spread of News on Digg and Twitter Social Networks. In: Proceedings of the 14th International AAAI Conference on Weblogs and Social Media, AAAI 2010. ACM, Atlanta (2010)
17. Romero, D.M., Meeder, B., Kleinberg, J.: Differences in the mechanics of information diffusion across topics: idioms, political hashtags, and complex contagion on Twitter. In: Proceedings of the 20th International Conference on World Wide Web, WWW 2011, Hyderabad, India, pp. 695–704 (2011)
18. Benevenuto, F., Rodrigues, T., Cha, M., Almeida, V.: Characterizing user behavior in online social networks. In: Proceedings of the 9th ACM SIGCOMM Conference on Internet Measurement Conference (2009)
19. Cheong, M., Lee, V.: A Study on Detecting Patterns in Twitter Intra-topic User and Message Clustering. In: 20th International Conference on Pattern Recognition, ICPR (2010)
20. Abel, F., Gao, Q., Houben, G.J., Tao, K.: Analyzing Temporal Dynamics in Twitter Profiles for Personalized Recommendations in the Social Web. In: Proceeding of the 3rd International Conference on Web Science, WebSci 2011 (2011)
21. Ruhela, A., Tripathy, R.M., Triukose, S., Ardon, S., Bagchi, A., Seth, A.: Towards the use of Online Social Networks for Efficient Internet Content Distribution. In: Proceedings of the Fifth International Conference on Advanced Networks and Telecommunication Systems, ANTS 2011. IEEE, Bangalore (2011)
22. OpenCalais: OpenCalais (2011), http://www.opencalais.com/ (accessed October 28, 2011)

Stylometric Analysis for Authorship Attribution on Twitter

Mudit Bhargava, Pulkit Mehndiratta, and Krishna Asawa

Jaypee Institute of Information Technology
muditbhargava09@gmail.com,
{pulkit.mehndiratta,krishna.asawa}@jiit.ac.in

Abstract. Authorship Attribution (AA), the science of inferring an author for a given piece of text based on its characteristics is a problem with a long history. In this paper, we study the problem of authorship attribution for forensic purposes and present machine learning techniques and stylometric features of the authors that enable authorship to be determined at rates significantly better than chance for texts of 140 characters or less. This analysis targets the micro-blogging site Twitter[1], where people share their interests and thoughts in form of short messages called "tweets". Millions of "tweets" are posted daily via this service and the possibility of sharing sensitive and illegitimate text cannot be ruled out. The technique discussed in this paper is a two stage process, where in the first stage, stylometric information is extracted from the collected dataset and in the second stage different classification algorithms are trained to predict authors of unseen text. The effort is towards maximizing the accuracy of predictions with optimum amount of data and users under consideration.

Keywords: Online Social Media, Twitter, Authorship Attribution, Machine Learning Classifier, Stylometry Analysis.

1 Introduction

In recent years, authorship attribution of anonymous messages has received notable attention in the cyber forensic and data mining communities. During the last two decades this technique has extended to computer facilitated communication or online documents (such as e-mails, SMS, Tweets, instant chat messages etc) for prosecuting terrorists, pedophiles, and scammers in the court of law. In the early 19th century it was considered difficult to determine the authorship of a document of fewer than 1000 words. The number decreased significantly and by the early 21st century it was considered possible to determine the authorship of a document in 250 words. The need for this ever decreasing limit is exemplified by the trend towards many shorter communications techniques like Twitter, Facebook[2], Short Message Services (SMS) etc.

---

[1] https://twitter.com/
[2] https://www.facebook.com/

V. Bhatnagar and S. Srinivasa (Eds.): BDA 2013, LNCS 8302, pp. 37–47, 2013.

Authorship Attribution of online documents is different from the authorship attribution of traditional work in two ways. Firstly, the online documents or text collection is mostly unstructured, informal and not necessarily grammatically correct as compared to literature, poems and phrases which are syntactically correct, very well structured and elaborative in nature. Secondly, for a single online document the number of authorship disputes are far more as compared to traditional published documents, that is because one of the challenges with authorship attribution in this case is scarcity of standardized data to test the accuracy of results.

Online social networks (OSN) like Twitter, Facebook, Linkedin[3] give a new dimension to the authorship attribution all together. These online networks provide effective and fast means of communication for the conduct of any criminal activity by anonymous users. User may use screen names or pen-names on these sites, while others may not provide the correct identification information with the accounts. On top of it, a single user can create multiple profiles on these online social networks. Anonymity poses bigger threats for law enforcement agencies in tracking the identity of these users. This paper is an effort towards addressing the problem of authorship attribution for an online social network Twitter. Twitter has surged popularity in recent years and now reports that it has over 500 million user base which share almost same number of messages (called *tweets*) per day [4].

This paper focuses on the problem of identification of the original author for a given tweet from a list of suspected authors for it using stylometric information. Various stylometric features have been taken into consideration for the training and later testing purposes of the machine learning algorithms such as Support Vector Machine (SVM) classifier. The experiments have been conducted on different datasets using the mentioned approach and in the last ideas for the future work have been discussed

2   Background

There has been an unprecedented growth in the amount of short messages that are shared worldwide everyday. Status messages on Twitter and Facebook, comments on the YouTube[4] and news blogs show a clear trend on using short messages for daily communication on the internet. With the advent of free text messaging mobile applications like Whatsapp[5], Viber[6] and the existing SMS, millions of short messages are shared through mobile phones.

A similar trend has also been seen in cyber-crime, where fraudulent activities like identity frauds and cyber-bullying are usually done with shorter messages such as fraud e-mails, forum posts, on Facebook and Twitter, as well as many other websites. Thus, it is of utmost importance to come up with some technique

---

[3] https://www.linkedin.com/
[4] https://www.youtube.com/
[5] http://www.whatsapp.com/
[6] http://www.viber.com/

or method to identify the authors of the various short messages posted daily on these websites. However, there are several stylometric features comprising of lexical, syntactic, structural, content-specific and idiosyncratic characteristics. Many studies have been proposed which consider stylometric features for performing the authorship attribution but still a lot of work has to be done for messages of length as low as 140 characters or less.

2.1 Related Work

In field of Stylometry, linguistic characteristics of a language are studied to gain knowledge about the author of the text. In [1], Abbasi *et al.* have talked about the issue that how the anonymity hinders the social accountability and tried to identify the author based on his/her writing style. Similarly in [2], authors have tried to apply mining techniques to identify the author of any particular e-mail. Koppel *et al.* [3] have tried to automatically categorize the written text and find out the gender of the author. Word frequencies, word length, number of sentences are considered important stylometric features as mentioned in [5].

The decrease in the size of the document has broadened the area of applications available. Early work on authorship attribution was focused on large documents like Federalist papers [8], but with recent developments in the field, attribution of authorship has even become applicable to blog posts [7] and short forum postings [6, 12].

R. Rousa Silva et.al [19] have worked on authorship attribution using stylistic markers for tweets written in Portuguese. Their analysis shows how emoticons and short messages specific features dominate over traditional stylometric features to determine the authorship of tweets. The final results show significant success (i.e. F Score = 0.63) for 100 examples available from each author under consideration. However the numbers of authors under suspect have been limited to 3 at a time.

Also, a recent technique involved using Probabilistic Context-Free Grammars (PCFG) for the attribution of author of a document [9]. Both lexical and syntactic characteristics were taken into consideration to capture an author's style of writing. The corpus that was included had data from different fields such as poetry, football, business, travel, and cricket. The method involved, first implementing a PCFG for all authors separately and then build a training model using the grammar for performing the classification. This model was combined with other models (bag-of-words Maximum Entropy classifier and n-gram language models) which captured lexical features too. For the dataset related to cricket, 95% of the total instances were correctly classified.

The next section provides an overview of the data set, how the tweets are collected, cleaned, pre-processed and clustered/grouped under one author. Section 4 discusses about the methodology used for the authorship attribution. In section 5 we discuss about the various results that we achieve including the accuracy of the experiments conducted. Finally, section 6 provides the conclusion of the experimental design used for the attribution of authors.

3   Data Set

Data always plays a critical role in authorship attribution. For performing the same on Twitter, an author attributed tweet dataset is required. Standard twitter corpora contain multiple tweets from multiple authors. Moreover, twitter terms of use do not allow distribution of tweets. Twitter corpora are collection of user IDs and tweet IDs. Downloading the content using automated scripts that accompany these corpora is a time consuming task, as twitter servers cannot be hit hard with too many requests at once. Owing to these constraints, we create our own dataset of author classified tweets using a Twitter client application that randomly collects public statuses using Twitter streaming application programming interface (API). A python based twitter corpus tool from [14] returns a random sample of about 5000 public statuses and stores them to disk in Java Script Object Notation (JSON) format.

Requirement is for the users/authors who would undergo the stylometric analysis. Choosing these users for experiments is a three step task. From the 5000 public statuses that have been collected, a list of unique authors is generated. Next, the requisite number of users are selected from the list randomly. Lastly, 300 most recent public statuses of these selected authors are streamed using GET statuses/user_timeline API from twitter [15].

We require users to have a certain threshold number of tweets (discussed in Section 4) and their language of profile and tweets to be English. Hence, if a selected user doesn't meet this criteria, we randomly select another user from the list of unique authors. Tweets streamed are parsed for their text or 'tweet' content and twitter specific features like 'hashtags', 'usermentions' and 'embedded urls'.The use and impact of these features is discussed in the further sections.

4   Methodology

Stylometric analysis on tweets is similar to those done on other forms of short texts such as web-forum posts or online instant messaging chats. They are informal and similar in structure and syntax [13]. An exhaustive feature set considering stylometric information is built for our experiment, however with an assumption that the authors unconsciously follow a specific pattern and are consistent in their choices [12]. Various broad categories of the features are as follows:

1. Lexical Features:
   (a) Total number of words per tweet
   (b) Total number of sentences
   (c) Total number of words per sentences
   (d) Frequency of dictionary words
   (e) Frequency of word extensions
   (f) Lexical diversity
   (g) Mimicry by length
   (h) Mimicry by word

Even though tweets are really short texts, users would manage to write them using dictionary words and framing proper sentences (features b, c, d). Feature (e) looks out for authors who would have a habit of extending the words by repetition of the last or intermediate letter, for example 'hiiii!!!', 'heyyy!', 'meee' etc. We learn from instant messaging [13] that how multiple messages from the same user are related in terms of vocabulary, length, attitude and that the user mimics his own style in each messages he sends. Though instant messages are different from tweets in numerous ways, we see quantifying mimicry as the ratio of length [13] in two chronological tweets (feature (g)) and as the number of common words in those two pieces of text, provide valuable information about the writing styles of the author. For example in the following tweets, 'Actually, I have to go out with friends today', 'Actually, I was a little busy today', 'Oh! Cool. That rocks!', 'Oh cool, the plan is on' we can see that how the author has a habit of using the words such as 'actually' or 'cool' quite often.

2. Syntactical Features:
   (a) Total number of beginning of sentences (BOS) characters capitalized
   (b) Number of punctuation per sentence
   (c) Frequency of words with all capital letters normalized over number of words
   (d) Frequency of alphanumeric words normalized over number of words
   (e) Number of special characters, digits, exclamation and question marks
   (f) No of upper case letters
   (g) Binary feature indicating use of quotations

There would be users ho would write their entire messages in capitalized letters, HELLO ARE YOU THRE? (feature (c)) and many users would condense words with the combination of characters and digits (like tonight becomes 2night or tomorrow becomes 2morrow, feature (d)) to fit in their entire content in the specified character limit. Features from (e)-(f) cover messages types where a different amount of special characters are used like "*, #, %, ^"etc., words are irregularly capitalized (eg. heLLo) or have too many question marks or exclamations (e.g. What??? or hi!!!). Feature (g) takes in account styles of those users who like to post popular quotes or quote other users.

3. Features specific to tweets:
   (a) Binary feature indicating if the tweet is a re-tweet
   (b) Number of hash-tags normalized over number of words
   (c) Number of user mentions normalized over number of words
   (d) Number of URLs normalized over number of words

There are features that are unique to twitter posts only. Re-tweet is sharing of a tweet originally composed by another author, hash-tags are used to convey the subject of a tweet (#sports,#now playing etc.), user mentions tag other users/sends them replies (@user1, @user2) and URLs are attached to share pictures, videos etc. Users would often bloat their tweets with hash-tags or user-mentions or have a very high frequency of re-tweeting. These features try and cover such stylometric information.

4. Other helpful features:
   (a) Frequency of Emojis[7]
   (b) Number of Emojis per word
   (c) Number of Emojis per character

Most emoticons (:), :P, :/) in data are converted to emojis and have a special unicode representation, making it difficult to detect them using syntactical features. Hence they need to be accounted for differently.

4.1   Grouping of Tweets

Even though a tweet can be at most 140 characters long, many authors use even lesser characters to express themselves. For example tweets like 'Good Morning followers' or replies like , '@someone Thanks!' are just 2-3 word long or 16-22 characters long. Such tweets would not have much stylometric information to contribute. One solution therefore is to remove such tweets from our dataset and work on texts with greater number of words. However in doing so, we loose out on important information about the style and traits of the author. Also if a user has a tendency of sharing only very short messages, with just 5-6 words, our system would fail to make predictions about such an author. Hence to overcome this challenge we group various tweets and increase the text size under consideration. Now to analyze patterns over a group of tweets rather than one single tweet is easier and fruitful. So we are mapping a group of tweets to its author rather than a single tweet and this is because some tweets might be excessively small as described above. The assumption here is that the tweets grouped together are from the same account and only one user maintains the twitter account under consideration. We also did a quantitative analysis, by bunching tweets in different group sizes. An overview of results obtained are summarized and illustrated in Table. 1 and Fig.1. Accuracy (%) is number of authors correctly classified. Group size greater than 4 provide acceptable results, but noticeably a group of 8-10 performs the best. A detailed analysis has been done in the further sections.

4.2   Experiments

As discussed in previous sections, a prerequisite to train machine learning algorithms and thereby performing text classification on the data collection is

---

[7] Japanese term for pictures characters and emoticons.

Table 1. Variation of accuracy with different tweet grouping sizes

Groupsize	Accuracy%
1	7.81
2	70.09
4	81.23
6	80.25
8	84.73
10	91.11

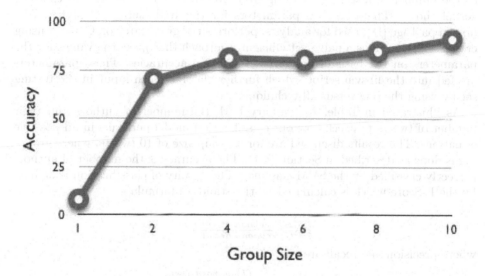

Fig. 1. Variation of accuracy with different tweet grouping sizes

selection of valid stylometric features. However these features need to be quantized for the collected data set. To perform that, a script has been generated that extracts features from the input tweets and groups them according to section (4.1) for further analysis. The script uses regular expressions and standard functions from the Natural Language Tool Kit (NLTK) [16] for python. The features and corresponding author labels will now be used to perform supervised machine learning using Support Vector Machine (SVM) [17].

4.3 Classification

Classification includes scaling of data, choosing the right SVM kernel, calculating best kernel parameter values and finally, testing the designed model for results. To avoid attributes with large numeric ranges to dominate over the ones with smaller ranges, it is essential to perform basic scaling of the attributes. So, if a feature has a value from [-10 to 10] it gets scaled to [-1 to 1]. Also, from various SVM kernels available it is important to choose the right one for best results. For this experiment, we pick the Radial Basis Function (RBF) kernel, which is a better choice over the linear, sigmoid or the polynomial kernels for several reasons in this scenario. RBF can handle cases where the relationship between class labels and features is non linear, this is because it non-linearly maps samples into a higher dimensional space. RBF also has fewer hyper-parameters affecting complexity of model selection and fewer numerical difficulties in comparison to polynomial and sigmoid kernels [10]. Since, features taken under consideration for this experiment range between [23-25] and are not very large in comparison to the number of instances (varying from 200 to a 1000), the RBF is the best kernel choice. There are two parameters for the RBF kernel: C and γ. The libsvm package [18], used for analysis, performs the grid-search on C and γ using cross validation. It is a naive yet efficient approach that picks the values for the parameters on the basis of best cross-validation accuracies. These parameters are fed into the libsvm script, which further classifies each input in the testing set by using the one versus all technique.

As illustrated in Table. 2 , we vary both the number of authors and their number of tweets to check how our classification model performs in all possible situations. The results discussed are for a group size of 10 tweets, where grouping is done as described in Section (4.1). The accuracy is the number of authors correctly classified by the SVM classifier. The quality of classification is defined by the F-Score which is calculated by the standard formula -

$$\frac{(2*Precision*Recall)}{(Precision+Recall)}$$

where precision and recall are as:

$$\text{Precision} = \frac{(True\ positives)}{(True\ positives + False\ positives)}$$

$$\text{Recall} = \frac{(True\ positives)}{(True\ positives + False\ negatives)}$$

The number of tweets varies from 200 - 300 and the number of authors varies from 10 - 20. The number of authors and tweets has been kept low with the following assumptions and constraints.

If we have fewer than 200 tweets, and we group them as explained earlier, we will end up with very few training instances per author. In a maximum grouping size of 10, we have 20 instances per author, which despite being less performs decently as it will be discussed in next section. If we have more than 300 tweets, we are asking for too much of data for one user, which might not always be available. Moreover, 300 most recent tweets give us sufficient stylometric information and we see in the next section that how the increase this number affects the performance of classification.

The number of authors have been kept low because, we consider that the list of suspected authors, who are under the scanner have been nailed down upon by other means like conduct of criminal/ non-criminal investigations. This is also the basis for the work done in [11] and [19].

Table 2. Results considering all features in a tweet grouping size of 10

Tweets	Users	Accuracy%	Precision%	Recall%	F-Score%
200	10	81.42	90.22	81.42	85.59
200	15	83.80	89.73	91.42	90.56
200	20	56.42	66.20	62.85	64.48
250	10	77.77	84.14	77.70	80.79
250	15	91.11	95.16	94.44	94.79
250	20	59.40	71.88	71.11	71.49
300	10	75.45	78.73	75.45	77.05
300	15	84.84	84.84	93.18	88.64
300	20	64.54	64.54	77.93	77.13

5   Results

Given just 200 tweets per author, and 10-15 suspects, we obtain an F-score in the range of (85.59% to 90.56%). However, if the number of suspects is increased to 20, the F-score drops drastically to a low value of 64.48%. Evidently in case of lesser number of tweets, we need to narrow down on our list of suspected authors. For 250 tweets per author, again best results have been achieved with 15 authors, where the F-score reaches 94.79%. With increase in data, we have an increase in F-Score for 20 authors (from 64.48% to 71.49%) indicating how an increase in content might be required with more number of users. With 300 tweets, again our best F-score is 88.64% with 15 authors. The F-score for 20 authors further increases with increase in number of tweets under consideration.

Table 3. Analysis of accuracy varying with features under consideration

Features	Accuracy%
All	91.11
Syntactica + Lexical + Tweet Specific	85.09
Syntactical + Lexical + Others	83.38
Syntactical + Tweet Specific + Others	83.38
Syntatical	72.20
Lexical + Tweet Specific + Others	62.74

There is notable decrease in the best obtainable accuracy as we increase the number of tweets. Since our tweets are collected in chronological order one possible inference from this observation is that the style of the author varies over time. Just similar to the case, when style of writing formal e-mails would differ from the ones to our friends and family, the style of writing tweets may differ over time because of the different topics that people are talking about over the time on Twitter. For example, if it's the Premier League season or there is an on-going cricket series, we would see a bias towards sports content amongst sport enthusiasts. This being a data dependent application, results are bound to vary over different data sets.

Table. 3 shows how different category of features make an impact on the accuracy of classification. Results are compiled for 250 tweets per author and 15 suspected authors. Non consideration of twitter specific features reduce the accuracy by 7.7%, however there is only a 6% decrease when features related to emoticons are eliminated from the analysis. Removal of syntactical features have a very strong impact on classification accuracy, reducing it by 28%. Though necessary, they are not sufficient for our experiment. Considering only syntactical features results in a low accuracy rate of 72.2 %.

As discussed in section 2, [19] also uses stylometric features for tweet authorship attribution. The former study requires 100 tweets per author and considers only 3 suspected authors at a time to achieve at most an F-Score of 0.63; by adding just a 100 more tweets our analysis can be extended to 10 suspected authors at a time and provide significantly better prediction accuracies (F-Score = 0.85, for 200 tweets, 10 users). With grouping of tweets in sets of 2-10 tweets per author and using the one versus all classification technique with SVMs, the techniques discussed in this paper and [19] are two different ways of using stylometric features for twitter authorship attribution; each having their own important contributions to the domain.

6   Conclusions and Future Work

Our study of authorship attribution for twitter shows interesting initial results. We have achieved a precision of up to 95.16 % , and a F-Score of up to 94.79% over a data set that is collected with no bias towards any specific content, user or geographical area. We also see how grouping of tweets together has an impact on author based tweet classification. It can be concluded that 200-300 tweets per author and list of 10-20 such suspected authors form a practical data set for analysis. The listed features, the SVM classifier with the RBF kernel and its optimum parameters values, form a good model for stylometric analysis of tweets from twitter.

In future, we would like to reduce the number of tweets per author required for defining a stylometric pattern. Also, a few points in the obtained results require detailed reasoning. These may become clearer by using an elaborate data set and performing further experiments. Other important tasks that we plan to undertake include, increasing the number of suspected authors under consideration and adding more precise features that would uniquely identify the tweets in question.

References

1. Abbasi, A., Chen, H.: Writeprints: A stylometric approach to identity-level identification and similarity detection in cyberspace. ACM Transactions on Information Systems (March 2008)

2. de Vel, O.: Mining e-mail authorship. In: ACM International Conference on Knowledge Discovery and Data Mining (KDD) (2000)
3. Koppel, M., Argamon, S., Shimoni, A.R.: Automatically categorizing written texts by author gender. Literary and Linguistic Computing 17(4), 401–412 (2002)
4. Twitter report twitter hits half a billion tweets a day (October 26, 2012),
   http://news.cnet.com/8301-1023_3-57541566-93/
   report-twitter-hits-half-a-billion-tweets-a-day/
5. Holmes, D.I.: The evolution of stylometry in humanities scholarship. Literary and Linguistic Computing 13(3), 111–117 (1998)
6. Abbasi, A., Chen, H.: Applying authorship analysis to extremist-group web forum messages. IEEE Intelligent Systems 20(5), 67–75 (2005)
7. Mohtasseb, H., Lincoln, U., Ahmed, A.: Mining Online Diaries for Blogger Identification. In: Proceedings of the World Congress on Engineering (2009)
8. Mosteller, F., Wallace, D.L.: Inference in an authorship problem. Journal of the American Statistical Association 58(302), 275–309 (1963)
9. Raghavan, S.: Authorship Attribution Using Probabilistic Context-Free Grammars. In: Proceedings of the 48th Annual Meeting of the Association for Computational Linguistics, ACL (2010)
10. Hsu, C.-W., Chang, C.-C., Lin, C.-J.: A Practical Guide to Support Vector Classification. Department of Computer Science, National Taiwan University, Taipei, Taiwan (2010)
11. Malcolm Walter Corney, Analysing E-mail Text Authorship for Forensic Purposes. Queensland University of Technology, Australia (2003)
12. Pillay, S.R., Solorio, T.: Authorship Attribution of web forum posts. APWG eCrime Researchers Summit (2010)
13. Cristani, M., Bazzani, L., Vinciarelli, A., Murin, V.: Conversationally-inspired Stylometric Features for Authorship Attribution in Instant Messaging. ACM Multimedia (October 29, 2012)
14. Twitter Corpus (2012), https://github.com/bwbaugh/twitter-corpus
15. Twitter (2013),
    https://dev.twitter.com/docs/api/1/get/statuses/user_timeline
16. Natural language Toolkit (2013), http://nltk.org/
17. Support Vector Machine (2000), http://www.support-vector.net/
18. Libsvm (2013), http://www.csie.ntu.edu.tw/cjlin/libsvm/
19. Sousa Silva, R., Laboreiro, G., Sarmento, L., Grant, T., Oliveira, E., Maia, B.: 'twazn me!!! ;(' Automatic Authorship Analysis of Micro-Blogging Messages. In: Muñoz, R., Montoyo, A., Métais, E. (eds.) NLDB 2011. LNCS, vol. 6716, pp. 161–168. Springer, Heidelberg (2011)

Copyright Infringement Detection of Music Videos on YouTube by Mining Video and Uploader Meta-data

Swati Agrawal and Ashish Sureka

Indraprastha Institute of Information Technology-Delhi (IIIT-D),
New Delhi, India
{swati1134,ashish}@iiitd.ac.in

Abstract. YouTube is one of the largest video sharing website on the Internet. Several music and record companies, artists and bands have official channels on YouTube (part of the music ecosystem of YouTube) to promote and monetize their music videos. YouTube consists of huge amount of copyright violated content including music videos (focus of the work presented in this paper) despite the fact that they have defined several policies and implemented measures to combat copyright violations of content. We present a method to automatically detect copyright violated videos by mining video as well as uploader meta-data. We propose a multi-step approach consisting of computing textual similarity between query video title and video search results, detecting useful linguistic markers (based on a pre-defined lexicon) in title and description, mining user profile data, analyzing the popularity of the uploader and the video to predict the category (original or copyright-violated) of the video. Our proposed solution approach is based on a rule-based classification framework. We validate our hypothesis by conducting a series of experiments on evaluation dataset acquired from YouTube. Empirical results indicate that the proposed approach is effective.

Keywords: YouTube Copyright Infringement Detection, Social Media Analytics, Mining User Generated Content, Information Retrieval, Rule-Based Classification.

1 Research Motivation and Aim

YouTube[1], Dailymotion[2] and Vimeo[3] are popular video sharing platforms on the Internet. YouTube is the largest and most popular free video-sharing service (having several social networking features in addition to video sharing) having millions of subscribers and users worldwide. YouTube allows users to upload and watch an unlimited number of videos online. It was started as a personal video sharing service and has evolved to a worldwide entertainment destination [5].

---

[1] http://www.youtube.com/
[2] http://www.dailymotion.com
[3] http://www.dailymotion.com

V. Bhatnagar and S. Srinivasa (Eds.): BDA 2013, LNCS 8302, pp. 48–67, 2013.
© Springer International Publishing Switzerland 2013

According to YouTube statistics[4]: over 6 billion hours of video are watched each month on YouTube and 100 hours of video are uploaded to YouTube every minute. YouTube suffers with several problems related to content pollution and misuse due to low publication barrier and anonymity. For example: presence of Spam and promotional videos [19], pornographic content [19] [20], hate and extremism promoting videos [21] [22], harassment and insulting videos [23] and copyright violated videos [1] [5] [14] [18]. Mining YouTube to understand content pollution and solutions to combat misuse is an area that has attracted several researcher's attention.

One of the major problems encountered by YouTube is uploading of copyright infringement videos (the focus of the work presented in this paper). Despite continuous attempts by YouTube to counter copyright infringement problem[5], the problem of copyright infringement still persists on a large scale and is a serious issue. Piracy in YouTube is found in television shows, music videos and movies. Copyright infringement costs the music industry and the government millions of dollars every year (the application domain presented in this paper is on copyright infringement detection of music videos). As per the Economic Times statistics (March 2013), the eventual increase in piracy on YouTube has affected the market value of right owners and over the past decade music and movie industry has faced a loss of 5000 crore rupees in terms of revenue (also a cost of 50000 jobs a year, has declined the theater collection from 95% to 60%[6].

The work presented in this paper is motivated by the need to develop a classifier to automatically detect copyright infringed videos on YouTube based upon a user centric query (filtering the YouTube search result for a given user query. We select an application domain (Hindi music and song videos) and observe that the search results on YouTube for several queries resulted in only a small percentage of original videos and majority of the search results consists of pirated or copyright infringed videos. We frame the problem as identification of original and copyright violated videos from the search results (for a query) targeted for the use-case of an end-user (content copyright owner and YouTube moderator) extracting original videos for a given query (or filter pirated videos).

The research objective of the work presented in this paper is the following:

1. To conduct a study (a case-study on music videos) on the extent of copyright infringement in YouTube and investigate solutions to automatically detect original and copyright infringement videos from the search results of a given user query.

2. To investigate the effectiveness of contextual features (video meta-data and not content) as discriminatory attributes for the task of original and copyright infringement video detection.

---

[4] http://www.youtube.com/yt/press/statistics.html
[5] http://www.youtube.com/t/contentid
[6] http://articles.economictimes.indiatimes.com/2013-03-05/news/37469729_1_upload-videos-youtube-piracy

2   Related Work and Research Contributions

2.1   Related Work

In this section, we discuss closely related work (to the study presented in this paper). We conduct a literature survey (refer list of related papers in Table 1) in the area of copyright infringement web video detection, duplicate and near-duplicate video identification. Based on our review of existing work, we observe that most of the researches and techniques for duplicate or copyright violated video detection are oriented towards the application of multimedia and image processing on the video content. The focus of our study is to investigate the effectiveness of meta-data (video contextual data) based features for the task of copyright infringement detection (which we see as relatively unexplored and a research gap) and is a different perspective to the problem than analyzing the video content.

YouTube itself uses some techniques to avoid and detect the copyright violated videos. It uses YouTube Help and Support Forums to make users aware about the Copyright infringement and uses a DMCA (Digital Millennium Copyright Act) to maintain an appropriate balance between the rights of copyright owners and the needs of users[7] [25]. YouTube also uses Content-ID System- a multi-media technique to detect copyright violated videos[9].

1. George H. Pike highlights some disused services (Napster and Grokster) and their users that have been extensively sued by US Govt. for their role in facilitating the exchange of copyrighted content. In his paper he mentions that even if YouTube has a different technical structure than Napster or Grokster, YouTube parallels the original Napster in its active role in the exchange of information between users. [1]
2. Xiao Wu et. al. outlines different ways to cluster and filter out the near duplicate videos based on text keywords and user supplied tags. They propose a hierarchical method to combine global signatures and local pairwise measure to detect clear near-duplicate videos with high confidence and filter out obvious dissimilar ones. [8]
3. Julien Law-To et. al. present a content based technique called ViCopT (Video Copy Tracking) for analyzing the behavior of local descriptors computed along the video. The proposed solution detects similar videos based upon the background content and watermarking scheme. It also gathers video tags and keywords metadata to find the similarity among videos. [14]
4. Lu Liu et. al. present a content based technique which detects real time duplicate videos by selecting video representative frame and reducing them into hash code values. Similarity of any two videos can be estimated by the proportion of their similar hash codes. [15]

---

[7] The Digital Millennium Copyright Act on YouTube is a law passed in 1998 governing copyright protection on the Internet under the second title i.e. 'Online Copyright Infringement Liability Limitation', section 512 of the Copyright Act to create new limitations on the liability for copyright infringement by on-line service providers[8]

[9] http://www.youtube.com/t/contentid

5. Hungsik Kim et. al. present Record Linkage Techniques and Multimedia based schemes to detect copied, altered or similar videos by comparing similarity among key-frame sequence, video signatures using MPLSH (Multi-Probe Locality Sensitive Hashing) techniques. [9]
6. Stefan Siersdorfer et. al. describe that content redundancy can provide useful information about connections between videos. Additional information obtained by automatically tagging can largely improve the automatic structuring and organization of content. [7]
7. Xiao Wu et. al. present a context and content based approach to detect real time near duplicate videos. The proposed method uses user-supplied titles, tags and other text descriptions and image thumbnail, caption, color, border of video to detect similar videos. [18]
8. Hyun-seok Min et. al. propose a semantic video signature, model-free semantic concept detection, and adaptive semantic distance measurement to detect similar videos. [16]

Table 1. Literature Survey of 8 Papers on Detecting Copyright Infringed and Content Redundant Videos using Data Mining and Multi-Media Techniques

Research Study	Objective
H. Pike. 2007 [1]	Describes some obsolete video hosting websites which were shut down due to the sharing of copyright content.
Xiao Wu et. al.; 2007 [8]	Proposed Video tags and keywords based analysis to detect Nearly duplicate videos.
Law-to et. al.; 2007 [14]	Proposed method is a content based approach that detects duplicate videos using temporal and ordinal features.
Lu Liu et. al.; 2007 [15]	A multi-media approach to detect real time duplicate videos by re-sampling the video frames and reducing their hash code values.
Kim et. al.; 2008 [9]	Proposed a Multimedia and hash based scheme to detect copied/altered/similar videos.
Siersdorfer et. al.; 2009 [7]	Assigning similar tags to different videos together based upon their content similarity.
Xiao Wu et. al.; 2009 [18]	Describes an approach to detect real time duplicate videos based using content and contextual features.
Min et. al.; 2012 [16]	A semantic concept based approach to detect similarity between two video shots.

2.2 Research Contributions

In context to existing work, the study presented in this paper makes the following unique contributions:

1. A solution for copyright infringement detection in YouTube using video meta-data. The proposed system is based on classifying each video (original or copyright violated) in YouTube search result (Top K) for a given user

query. The proposed framework and setting of identification of copyright infringement and original video from the YouTube search results for a given
user query is novel in context to existing work.
2. An empirical analysis (on an experimental dataset and a specific domain) on
    the technical challenges and effectiveness of various video contextual features
    for the copyright detection task using video meta-data. The discriminatory
    features for classification and underlying algorithm are a fresh perspective
    to traditional techniques.

3   Technical Challenges

Automatic detection of copyright violated video in the search results for a given
query is a non-trivial and a technically challenging problem because of the following three reasons.

3.1   Arbitrary Number of Irrelevant Videos in Search Results

We define relevant video in the search result for a given query if the video (irrespective of whether it is copyright infringement or original) belongs to the same
song or music as the query. A video is defined as irrelevant if it is not the actual
video the user is looking for but it is retrieved into the search result when queried.
For example, a music video from the same album or movie or a non-music video
with some common words as query in title. We observe that YouTube search
returns several irrelevant results for a given music title and hence identifying
relevant videos (or filtering irrelevant videos) is necessary before detecting if the
video in the search result is original or copyright violated (as it does not matter
if a video is original or copyright violated if it does not match the information
need and query of the user). We notice that the number of irrelevant videos
returned in search result (for a given query) varies which makes the problem
of recognizing relevant videos (before recognizing original or copyright violated)
technically challenging. We conduct an experiment to demonstrate the technical
challenge by creating a dataset of 50 queries (song titles) and top 20 videos in
the search result. We manually label each search result as relevant or irrelevant.
Among these top 20 search results, we count the number of relevant videos (R)
and irrelevant videos (20 - R) for each query. Figure 1a shows the histogram
for the number of irrelevant videos for 50 queries. It reveals that the number of
irrelevant videos varies from 0 to 15. We observe that for some queries 75% of
the search result consisted of irrelevant videos. The presence of irrelevant videos
and wide variation in the number of irrelevant videos within Top K search result
poses technical challenges to the relevant or irrelevant video classifier which is
a step before the original or copyright-violated video classifier (See Figure 2).
Figure 1a illustrates that for some queries (Query ID 5, 16, 18, 36, 38, 50) there
are no irrelevant videos and for queries (15, 35) 75% of the search result are
irrelevant videos.

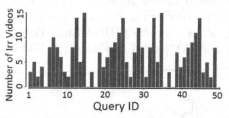 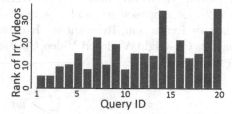

(a) Arbitrary Number of Irrelevant Videos in Search Results

(b) Non-Uniform Ordering of Irrelevant Videos

Fig. 1. Variations in Number of Irrelevant Videos and their Rank for Each Query, Irr=Irrelevant

3.2 Non-uniform Ordering of Irrelevant Videos

We annotate the irrelevant videos in our experimental dataset (50 queries of song title and 20 search result) with their rank in the Top 20 search result. Figure 1b displays the distribution of the number of irrelevant videos across 20 ranks (Rank 1 to 20). We observe that while irrelevant video are present at Rank 1 up-to Rank 20. Figure 1b reveals that there is a wide variation in the rank of irrelevant videos which poses challenges to the relevant or irrelevant video classifier. Our experiment reveals that the number of irrelevant videos and their rank within the Top K search result varies across queries and makes the problem of relevant (to the given query) and original video detection challenging. In our experimental dataset, for 5 queries irrelevant videos retrieved on 1st rank and for 22 queries irrelevant videos were on 9th rank. Our experiment demonstrates that irrelevant videos are ranked high in the search result for several queries.

3.3 Noisy Data in Video Contextual Information

Presence of low quality content, Spam, and noise is prevalent in popular social media websites (including YouTube) due to low publication barrier and anonymity. We observe that YouTube video metadata such as title and description contains several spelling and grammatical mistakes and presence of short-forms and abbreviations. We also notice missing (for example, absence of video description) and misleading or incorrect information (for example, a promotional video containing title or description or tags of a popular video). Noise, missing and incorrect information in YouTube meta-data poses technical challenges (text processing and linguistic analysis) to the classifier based on mining contextual information on YouTube Videos.

4   Solution Design and Architecture

Figure 2 presents a general framework for the proposed solution approach. We divide copyright infringement detection problem into sub-problems: detecting irrelevant videos and detecting copyright violated videos. As shown in the Figure 2, the proposed method is a multi-step process primarily consists of Test Dataset Extraction, Relevant vs. Irrelevant Video Classifier (IRVC) and Original vs. Copyright Violated Video Classifier (OCIVC) cited as Phase 1, 2 and 3 respectively.

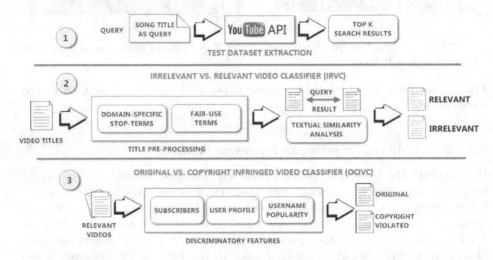

Fig. 2. A general framework for our proposed solution approach

4.1   Test Dataset Extraction

During our study, we first conduct a manual analysis and visual inspection of search results (consisting of copyright violated and original YouTube videos) for several queries consisting of music video titles. We define and propose a solution approach based on our manual and empirical analysis of a sample data. Once we define a solution approach, we create an experimental test-bed to evaluate the effectiveness of our approach (the Phase labeled as Test Dataset Extraction). The test dataset consists of 100 song titles (collected from 20 users). We retrieve Top 20 (K = 20 in our case-study) search results for each video (in the test dataset) using the YouTube Java API[10]

4.2   Irrelevant vs. Relevant Video Classifier (IRVC)

The input test dataset for this phase is the title of videos in search result extracted from Phase 1. IRVC is a binary classification scheme to filter irrelevant and relevant video in the test dataset. A video is said to be irrelevant, if the video does not belong to the same song or music as the query. For example a non-music video or different song from same album with some common words as query in title. As shown in the Figure 2 IRVC is further divided into three modules i.e. Title Pre-Processing, Textual Similarity and Classification.

---

[10] https://developers.google.com/youtube/2.0/developers_guide_java

4.2.1   Title Pre-processing

In this phase of the pipeline, we pre-process the title to remove pre-defined terms. This phase uses three basic linguistic discriminatory features for pre-processing or tokenization: Query and title of the search result, lexicons of Domain Specific Stop Words (DS2W) and Fair-Use Keywords (FKW). Stop words which are common and non-content bearing and which are not useful for finding relevance of a video to a given query are removed their presence in the title negatively impacts the performance of the algorithm. To perform tokenization, we create our own domain specific stop-words (these stop-words are specific to the case-study or domain presented in this paper) which is used as a lexicon by the text-processing algorithm. We identify stop-words like: bluray, HD, must watch and 1080p. Fair-Use of a video is a lawful copyright infringement permitting limited use of copyrighted material without acquiring permission from the right holders or the content owner. We define a video to be a fair-used video, if any keyword from FKW is present only in the video title but not in the query. For example, if the user queried for a song on YouTube: cover, flute, instrumental version of the video (if present) are also retrieved in the search results (by the YouTube search engine). However, these video are considered to be as Fair-Use Videos by our algorithm (as it is a limited use of the copyright content without violating the copyright terms). Based upon the annotator's decision, we filter fair-used videos as irrelevant videos because we believe the user (who searched for the query) does not intend to watch these videos (otherwise it would have been explicitly mentioned in the query). We look for these fair use keywords in each search titles and for each video assign a score of either 0 or 1 after this lookup (0 represents the video to be a fair use video and 1 represent it to be a non-fair use video).

4.2.2   Textual Similarity

In this sub-phase, we find the extent of similarity between the query and title of a video in search result. The textual similarity is being performed on three levels. First, we compare the query and video title from search result on statement level. If the query is a substring of the title then we call it a match, otherwise we perform a word level comparison to find the percentage of a video title that matches with query. To avoid the noisy data we perform this comparison at character level for each word. A video is said to be relevant if the video title matches with the query scoring above a specified threshold. After textual similarity comparison each video in search result gets assigned score between 0 and 1; where 0 represents the no matching and 1 represents the perfect matching.

4.2.3   Classification

In the final sub-phase of IRVC, the input is an array of all video titles in search results and their respective scores in previous sub-phases of IRVC. The aim of this sub-phase is to classify relevant and irrelevant videos based upon the scores of each video in previous sub-phases. We used a Rule Based Classification scheme to classify these videos. Each sub-phase has it's own specific threshold value. If the video in the search result, scores above the specified threshold values in respective sub-phases, is classified as relevant otherwise irrelevant.

4.3   Original vs. Copyright Infringed Videos Classifier (OCIVC)

A video is defined as Copyright Infringed video, if it is shared or re-produced illegally without the permission and knowledge of the copyright holder. In terms of YouTube, a video is said to be copyright violated if the uploader of the video is unauthorized to upload it on web[11]. The aim of this phase is to classify such violated videos from original videos in our test dataset. The input test dataset for this phase is the Video IDs of all relevant videos classified by IRVC in Phase 2. We searched for few official and unofficial channels (Hindi Music Video Channels only) on YouTube to analyze the behavior of uploader metadata. To classify original and copyright violated videos we used three discriminatory features: Number of Subscribers, Number of Google Page Hits and YouTube Profile Information of uploader. These features are the quantitative measures used in the classification process and are selected by applying in-depth manual analysis on several user channels and their metadata. We obtained these features using YouTube and Google custom search API.

4.3.1   Discriminatory Features
Fig 3a and 3b illustrate that the number of subscribers and Google hit counts for original channels are very large in comparison to the violated channels. X-axis shows the YouTube User IDs (1 to 20 as Copyright Violated and 21 to 40 as Original Channels)[12]. Y-axis represent the value of Subscribers (Fig 3a) and Google Hit Counts (Fig 3b). All values on Y axis are normalized between 0 and 1. For each video, the computed score for these features are their value itself. Figure 3c shows the profile information of three different YouTube channels. This feature identifies only violated channel based upon some keywords. For example uploader profile having keywords as Fanpage, Fan made, Unofficial etc. As shown in the Fig 3c all three channels are created with the name of same artist but only the third channel[13] is a legitimate channel (verified by the artist himself). We can see that first two channels ('AkshayKumar WestBengalFans'[14] and 'AkshayKumarVEVO'[15]) have some keywords in their profile information 'fan club' and 'unofficial' which proves those channels to be violated channels. But The third channel has no profile information so we can not classify it in either of the category. Also we can not use some keywords to classify an original

---

[11] The three exclusive rights under U.S. copyright law that are most relevant to YouTube are the rights "to reproduce the copyrighted work", "to perform the copyrighted work publicly" and "to display the copyrighted work publicly".

[12] With the best of our knowledge, these official user channels are verified from their official websites and all feature values are correct as per March 2013 statistics.

[13] www.youtube.com/user/akshaykumar

[14] www.youtube.com/user/AkshayKumarWestBengalFans

[15] www.youtube.com/user/AkshayKumarVEVO

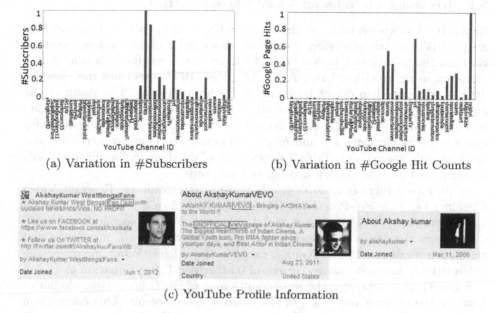

(a) Variation in #Subscribers   (b) Variation in #Google Hit Counts

(c) YouTube Profile Information

Fig. 3. Behaviour of Discriminatory Features for Original and Violated User Channels, # = Number of

channel because those words can be used by any user in their profile information. For example, in Figure 3c Channel 2 is a violated channel despite of being a "VEVO" channel[16].

4.3.2   Classification

The aim of this sub-phase is to classify original and copyright infringed videos based on their value for all discriminatory features. While studying the behavior of uploader metadata we created a small lexicon of a few official channels (verified from the official websites of respective music and movie production company). If the uploader of a video exists in that lexicon we classify that video to be an original otherwise for the rest of the videos we classify them using Rule Based Classification technique where each discriminatory feature has it's own specific threshold value.

5   Solution Implementation

In the Section, we present solution implementation details for the design and architecture articulated in the previous section.

---

[16] VEVO is a joint venture music video website operated by Sony Music Entertainment, Universal Music Group, and Abu Dhabi Media with EMI licensing its content to the group without taking an ownership stake. Source: http://en.wikipedia.org/wiki/Vevo

5.1   Irrelevant vs. Relevant Video Classifier (IRVC)

In binary classification problem, the class label is either positive or negative. The goal of IRVC is to classify relevant (positive class) and irrelevant videos (negative class) in the test dataset (video titles of top 20 search results for a song query, retrieved using YouTube JAVA API). Algorithm IRVC describes the classifier we have developed to detect irrelevant and relevant videos. The result of the algorithm shows the video IDs V' of all relevant videos. Input to the algorithm is a user's query (song title) q, an array V of video IDs of all search results for query q and two lexicons D_{st} and Key_{fair} of domain specific stopwords and fair use keywords respectively. Table 2 shows some of the domain specific stopwords we have used for tokenization and fair use keywords.

Steps 1 to 9 represent the first sub-phase of Irrelevant vs. Relevant Video Classifier i.e. Title Pre-Processing (Figure 2). Steps 2 to 5 removes all domain specific stopwords D_{st} from query q and video title t in the search result. Steps 6 to 9 finds the existence of specific fair use keywords Key_{fair} in query q and the search result t and if the Key_{fair} exists in the video title but the query, the video titles is assigned a score, $Score_1$ of 0 otherwise 1. See Equation 1.

Steps 11 to 19 represent the second sub-phase of IRVC i.e. textual similarity between the query q and title t of the video in search result. This comparison is being performed on three levels: computing the similarity on statement level and if the query is not a substring of the video title then we find the total number of words in query matching with the video title and a character level comparison to avoid noisy data using Jaro-Winkler Distance[17]. If the query is a perfect substring of the video title then we assign it a score, $Score_2$ equal to the query length $qlen$, otherwise we find the maximum number of words of query matching with the title stated as *match* array in step 13. These number of words are assigned as $Score_2$ to the respective video titles. See Equation 3. Threshold values to decide the similarity at word and character level comparison are shown in Equation 2. Steps 20 to 23 represent the classification procedure and labeling of videos as relevant or irrelevant depending upon the threshold measures.

$$Score_1 = \left\{ \begin{array}{ll} 0, & if \ \exists Key_{fair} \in t \wedge \forall Key_{fair} \notin q \\ 1, & otherwise \end{array} \right\} \tag{1}$$

$$\boxed{t_{NumWord} = 0.7 \text{ and } t_{NumChar} = qlen * 0.6} \tag{2}$$

$$\boxed{Score_2 = match.length(); \text{ where match} = qword \cap tword} \tag{3}$$

5.2   Original vs. Copyright Infringed Video Classifier (OCIVC)

OCIVC is a binary classification algorithm to classify copyright violated (positive class) and original (negative class) videos in the test dataset (Relevant Videos

---

[17] http://en.wikipedia.org/wiki/Jaro%E2%80%93Winkler_distance

---

Algorithm 1. Relevant vs Irrelevant Video Classifier

Data: Domain Specific Stopwords $D_{st} \in LexiconL1$, Fair-Use Keywords
    $Key_{fair} \in L2$, User Query q, Video IDs $V_{id} \in V$, $t_{NumChar}$, $t_{NumWords}$
Result: V', Video IDs of Relevant Videos

1   $V_{title} \leftarrow$ V.title(); $qlen \leftarrow$ q.length();
2   **for** *all* $t \in V_{title}$, q **do**
3   **if** *($\exists D_{st} \in t$)$\vee$($\exists D_{st} \in q$)* **then**
4       $t \leftarrow$ t.remove(D_{st});
5       $q \leftarrow$ q.remove(D_{st});

6   **for** *all* $t \in V_{title}$ **do**
7   **if** *(($\exists Key_{fair} \in t$)$\wedge$($\nexists Key_{fair} \in q$))* **then**
8       $Score_1 \leftarrow 0$;
    else
9       $Score_1 \leftarrow 1$;

10   **for** *all* $t \in V_{title}$, q **do**
11   **if** *(t.contains(q))* **then**
12       $Score_2 \leftarrow$ qlen;
    else
13       $match \leftarrow$ qword$\cap$tword;
14   **for** *all* $words \in$ qword $\setminus$ *tword* **do**
15         $flag \leftarrow JaroWinkler$(qword, *tword*);
16   **if** *(flag $\geq t_{NumChar}$)* **then**
17          $match \leftarrow$ match.add(qword);

18       $mlen \leftarrow$ match.length();
19       $Score_2 \leftarrow$ mlen;

20   **if** *($Score_1 \leq 0$)$\vee$($Score_2 \leq t_{NumWords}$)* **then**
21       $Class \leftarrow$ Irrelevant;
    else
22       $Class \leftarrow$ Relevant;
23       $V' \leftarrow V_{id}$;

24 **return** V';

Table 2. A Sample of Few Keywords from Domain Specific Stopwords and Fair-Use Video Lexicons

Domain Specific Stopwords					Fair-Use Keywords				
dvd	eng	latest	1080p	version	vm	live	cover	piano	karaoke
mp4	hindi	subtitle	bluray	exclusive	dj	award	flute	teaser	fanmade
avi	episode	youtube	trailer	must watch	iifa	remix	chorus	perform	instrumental

Classified by IRVC). Algorithm 2 describes proposed solution approach for Original vs. Copyright Infringed Video Classifier. The algorithm returns video IDs of all violated videos based on based on the application of discriminatory features.

The input test dataset for this classifier is the video IDs of all relevant videos classified by IRVC referred as V'. The algorithm takes the input of a lexicon of some domain specific official channels ($User_{off}$) on YouTube and threshold values for all discriminatory features. Table 3 shows some of these official channels (channels are verified from their official websites). Steps 1 to 3 extract uploader ids U_{id} (YouTube user channel name) of all relevant videos V' (classified by Algorithm IRVC) and store them in an array U'. Steps 4 to 7 extract all required metadata of each U_{id} i.e. Number of Subscribers, Number of Google Page Hits and User Profile Information represented as N_{subs}, $N_{pagehits}$ and profile respectively. These feature values are retrieved using YouTube[18] and Google Custom Search[19] APIs.

We perform a manual inspection on several official and unofficial channels (related to Hindi music videos) and by analyzing the behavior of discriminatory features for original and violated channels, we compute a threshold value for each feature. Steps 8 to 15 perform the Rule Based Classification using these feature values. The algorithm classifies a video as original if the feature values satisfy the threshold measures. Equations 4 and 5 show the threshold computation for N_{subs} and $N_{pagehits}$ respectively. In steps 8 to 10, we look up for some keywords in the user profile information; for example fan-page, fan-based, unofficial etc. If the profile information of the uploader contains any of these keywords, the video is labeled as copyright infringed video. For rest of the video we used Rule Based Classification scheme by comparing the values of N_{subs} and $N_{pagehits}$ with their respective threshold values, shown in Equation 6.

$$t_{sub} = \frac{\sum_{i=1}^{n} User_{off_i} . subscribers}{n} \tag{4}$$

$$t_{hitcount} = \frac{\sum_{i=1}^{n} User_{off_i} . pagehits}{n} \tag{5}$$

$$t_{sub} = 6415 \text{ and } t_{hitcount} = 1051 \tag{6}$$

Table 3. A Sample of Few Domain Specific Official Channels on YouTube (Hindi Music and Movies Production Companies)

Official YouTube Channels				
yrf	rajshri	ZoomDekho	tipsfilms	tseries
saavn	UTVGroup	sonymusicindiasme	ErosNow	venusmovies
T-Series	etcnetworks	Viacom18Movies	muktaart	shemarooent
starone	channelvindia	shemaroomovies	channelvindia	TheBollywoodShow

---

[18] https://developers.google.com/youtube/
[19] https://developers.google.com/custom-search/v1/overview

Algorithm 2. Original vs Copyright Infringed Video Classifier

Data: Video ID $V'_{id} \in V'$, official YT channels $User_{off} \in LexiconL3$, User
profile keywords $Key_{unoff} \in L4$, Subscriber Count threshold t_{sub},
pagehit threshold $t_{hitcount}$

Result: V''_t, Title of Infringed Videos

```
1  for all V'_id ∈ V' do
2  │   U_id ←UploaderName of video;
3  └   U' ←U'.add(U_id);

4  for all U_id ∈ U' do
5  │   N_subs ←SubscriberCounts();
6  │   N_pagehits ←SearchPageCounts;
7  └   Profile ←UserInfo;

8  for all N_subs, N_pagehits, Profile do
9  │   if (∃Key_unoff ∈Profile) then
10 │   │   Class ←Violated;
11 │   └   V_t" ←Title of V'_id;
12 │   if (∃User_off ∈U')∨((N_subs ≥ t_sub)∧(N_pagehits ≥ t_hitcount)) then
13 │   │   Class ←Original;
   │   else
14 │   │   Class ←Violated;
15 │   └   V_t" ←Title of V'_id;

16 return V_t";
```

---

6 Empirical Analysis and Performance Evaluation

6.1 Experimental Dataset

We asked 20 graduate students (to remove author or one-person bias) from our Department to help us create and annotate test dataset. We asked each student to provide 5 search queries (of Hindi song titles). We made sure that there are no duplicate queries (removing duplicate and requesting for more queries) and all 100 search queries (5 query each from 20 students) are unique. We searched and retrieved Top 20 results from YouTube for each of the 100 queries (using YouTube Java API). We then annotated each search result (a total of 2000 search results) with the help of the students from whom we collected the queries. Each search result was labeled as Relevant or Irrelevant and Original or Copyright Violated. Manual labeling of videos revealed that there were 1490 relevant and 510 irrelevant videos. OCIVC takes an input of 1684 video search results and based upon the manual analysis, we revealed that there were 1533 copyright violated and 151 original videos. Our goal is to classify original and violated videos among all relevant videos in test dataset. Based upon our analysis we label 1413 videos as relevant & original and 121 videos as relevant & copyright violated.

6.2   Performance Metrics

To measure the effectiveness of our solution approach we used standard confusion matrix. For a binary classifier each instance can only be assigned to one of the classes: Positive or Negative. We evaluate the accuracy of our classifier by comparing the predicted class (each column in the matrix) against the actual class (each row in the matrix) of the video search results in the test dataset. The performance of the classifier is computed in terms of Sensitivity and Specificity by computing TPR, TNR, FPR, FNR, Precision, Recall, and F-score. TPR is the proportion of actual positives, which are predicted positive. TNR is proportion of actual negative, which are predicted negative. Precision is the proportion of predicted positives, which are actually positives, and Recall is the proportion of actual positives, which are predicted positive. F-Score is the weighted harmonic mean between Precision and Recall. F-score is being computed by assigning the equal weights to both Precision and Recall. The relation between Sensitivity and Specificity is represented in a ROC curve (FPR on X-axis and TPR on Y-axis). Each point on the ROC curve represents a sensitivity/specificity pair.

6.3   Experimental Results

6.3.1   Irrelevant VS. Relevant Video (IRVC) Classifier

Table 4 shows the confusion matrix for IRVC. Given the input of 2000 video search results, IRVC classifies 1208 (1119 + 89) videos as relevant and 792 (371 + 421) as irrelevant videos. There is a misclassification of 24.9% and 17.45% in predicting the relevant and irrelevant videos respectively. This misclassification occurs because of the noisy data such as lack of information, ambiguity in songs titles or misleading information.

Table 4. Confusion Matrix for IRVC

		Predicted	
		Relevant	Irrelevant
Actual	Relevant	75.10% (1119/1490)	24.90% (371/1490)
	Irrelevant	17.45% (89/510)	82.55% (421/510)

6.3.2   Original VS. Copyright Infringed Video Classifier (OCIVC)

Table 5 illustrates the confusion matrix for OCIVC. The classifier takes the input of 1684 (2000-316 Fair Use Videos) video search results classifying 1469 (1455 + 14) videos as copyright violated and 215 (78 + 137) videos as original. Considering the manual labeling on each search result there is a misclassification of 5.1% and 9.27% in predicting the copyright violated and original videos respectively.

As we mentioned above we are using popularity measures (Number of Subscriptions and Number of Google Page Hits) of a YouTube user account (uploader of the video) to detect is as violated or original. Many users name their channels similar to the official channels For example, KingKhanHD[20]. Many such channels upload a large number of HD videos and get equal number of subscribers as the original channels. For example, YouTube has an official channel of T-Series Music Company i.e. 'tseries' and a channel named 'TseriesOfficialChannel' is created by a violated uploader to gain the attention of other YouTube users and to get a large number of subscribers.

Table 5. Confusion Matrix for OCIVC

		Predicted	
		Violated	Original
Actual	Violated	94.90% (1455/1533)	5.10% (78/1533)
	Original	9.27% (14/151)	90.73% (137/151)

6.3.3 Combined Classifier

Table 6 shows the confusion matrix for the combined classifier (combining IRVC and OCIVC in pipelining). The classifier takes the input of 2000 video search results classifying 982 (919 + 63) videos as relevant as well as copyright violated and 552 (494 + 58) videos as relevant as well as original. The accuracy of this classifier is dependent on the number of relevant & original videos and relevant & copyright violated videos. Therefore based upon the manual analysis on each search result there is a misclassification of 33.76% and 52% in predicting the copyright violated and original videos respectively.

Table 6. Confusion Matrix for Combined Classifier

		Predicted	
		Violated	Original
Actual	Violated	66.24% (919/1413)	33.76% (494/1413)
	Original	52.00% (63/121)	48.00% (58/121)

Table 7. Performance Results for Test Dataset for IRVC, OCIVC and Combined Approach classifiers. TPR=True Positive Rate, FNR= False Negative Rate, FPR= False Positive Rate, TNR= True Negative Rate, Acc= Accuracy, Prec=Precision and CA= Combined Approach

Classifiers	TPR	FNR	FPR	TNR	Acc	Prec	Recall	F-Score
IRVC	0.77	0.23	0.20	0.80	76.80%	81%	75%	0.78
OCIVC	0.95	0.05	0.09	0.91	94.68%	91%	95%	0.93
CA	0.66	0.34	0.52	0.48	65.00%	56%	66%	0.61

---

[20] http://www.youtube.com/user/KingKhanHD

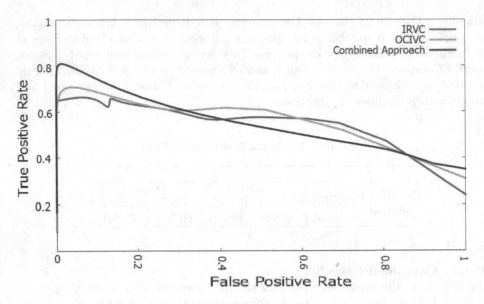

Fig. 4. ROC Curve for IRVC, OCIVC and Combined Classifiers

The accuracy of combined classifiers depends on the accuracy of both IRVC and OCIVC. In IRVC the misclassification of 24.9% affects the accuracy results of OCIVC and so do the combined classifier. Because in combined classifier OCIVC receives only relevant videos filtered by IRVC. Table 7 shows the accuracy results (Accuracy, Precision, Recall and F-Score) for IRVC, OCIVC and combined classifiers.

Figure 4 shows the ROC Curve for all three classifiers. This curve plots the True Positive Rate (Y-axis) against the False Positive Rate (X-axis) for all video search results in test dataset. As the graph illustrates that the curve for combined approach is below than OCIVC and has less accuracy. But the area under curve is large enough, hence we can say that experimental tests performs well. The area between 0.60 and 0.80 shows that the tests are upto a reasonable accuracy.

7   Limitations and Threats to Validity

We conduct experiments on one application domain and a test dataset consisting of 100 queries and 2000 videos (in search result). Conducting similar experiments on a larger dataset and multiple domains can increase the generalizability of the result. The sample queries are provided and test dataset (irrelevant and relevant videos, original and copyright violated) is annotated by graduate students at our University which can be biased to the annotators decision (we did find some disagreement between annotators on relevant or irrelevant video tagging). The lexicon is manually created (a tedious task and non-exhaustive) and an automatic way of constructing a lexicon will enhance the domain transferability of the solution approach.

8 Conclusion

We present an approach (based on mining contextual features or meta-data of a video) to detect original and copyright violated videos on YouTube for a given user query (tagging search results as original or copyright infringed). Proposed method is based on a rule-based (derived from empirical analysis of the dataset) classification framework. Experimental results on the test dataset demonstrate that the proposed approach (a multi-step process consisting of two classifier in pipeline) is able to discriminate between original and violated video with an accuracy of around 60% (above 75% accuracy for each of the two classifiers as part of the ensemble). Based on our analysis of contextual features, we conclude that the number of subscribers and page-hit counts (of the video or uploader id as a proxy to popularity) can be used as discriminatory features. We conclude that the profile information of the uploader is useful to detect violated channels but not a reliable indicator for identifying original channels.

Acknowledgement. The Authors would like to acknowledge Annapurna Samantaray for her help in evaluation phase of the work. The work presented in this paper is supported by the DIT (Department of Information Technology, India), research project grant of IIIT-D.

References

1. Pike, G.H.: Legal Issues: Google YouTube Copyright and Privacy. Information Today 24(4), 15 (2007)
2. Library of Congress, How to Investigate the Copyright Status of a Work, United States copyright office, Washington, DC 20559 (January 1991)
3. Russ Versteeg Viacom V/S YouTube: Preliminary Observations North Carolina. Journal of Law & Technology 9(1) (Fall 2007)
4. Kim, E.C.: YouTube: Testing the Safe, Harbors Of Digital, Copyright Law 17 S. Cal. Interdisc. L.J. 139 (2007-2008)
5. Breen, J.C.: YouTube or YouLose? Can YouTube Survive a Copyright Infringement Lawsuits. UCLA School of Law Year, Texas Intellectual Property. Journal 16(1), 151–182 (2007), http://works.bepress.com/jasonbreen/1
6. Wang, A.L.-C.: An Industrial-Strength Audio Search Algorithm. In: Choudhury, S., Manus, S. (eds.) The International Society for Music Information Retrieval, 4th Symposium Conference on Music Information Retrieval, ISMIR 2003, pp. 7–13 (October 2003), http://www.ismir.net, http://www.ee.columbia.edu/~dpwe/papers/Wang03-shazam.pdf
7. Siersdorfer, S., Pedro, J.S., Sanderson, M.: Automatic Video Tagging using Content Redundancy. In: 32nd International ACM SIGIR Conference on Research and Development in Information Retrieval, pp. 395–402 (July 2009)

8. Wu, X., Hauptmann, A.G., Ngo, C.-W.: Practical Elimination of Near-Duplicates from Web Video Search. In: Proceedings of the 15 International Conference on Multimedia, pp. 218–227. ACM, New York (2007)

9. Kim, H., Lee, J., Liu, H., Lee, D.: Video Linkage: Group Based Copied Video Detection. In: Proceedings of the 2008 International Conference on Content-Based Image and Video Retrieval, CIVR 2008, pp. 397–406 (July 2008)

10. Paisitkriangkrai, S., Mei, T., Zhang, J., Hua, X.-S.: Scalable Clip-based Near-duplicate Video Detection with Ordinal Measure. In: Proceedings of the ACM International Conference on Image and Video Retrieval, CIVR 2010, pp. 121–128 (2010)

11. Shen, J., Mei, T., Gao, X.: Automatic Video Archaeology: Tracing Your Online Videos. In: Proceedings of the Second ACM SIGMM Workshop on Social Media, WSM 2010, pp. 59–64 (2010)

12. Zhu, G., Yang, M., Yu, K., Xu, W., Gong, Y.: Detecting Video Events Based on Action Recognition in Complex Scenes Using Spatio-Temporal Descriptor. In: Proceedings of the 17th ACM International Conference on Multimedia, pp. 165–174 (October 2009)

13. Cohen, W.W., Ravikumar, P., Fienberg, S.E.: A Comparison of String Distance Metrics for Name-Matching Tasks. In: Proceedings of IJCAI 2003 Workshop on Information Integration, pp. 73–78 (August 2003)

14. Law-to, J., Buisson, O., Chen, L., Ipswich, M.H., Gouet-brunet, V., Joly, A., Boujemaa, N., Laptev, I., Stentiford, F., Ipswich, M.H.: Video copy detection: a comparative study. In: CIVR, pp. 371–378 (2007)

15. Liu, L., Lai, W., Hua, X.-S., Yang, S.-Q.: On Real-Time Detecting Duplicate Web Videos. In: IEEE International Conference on Acoustics, Speech and Signal Processing, ICASSP 2007, vol. 1, pp. 973–976 (2007) ISSN 1520-6149

16. Min, H.-S., Choi, J.Y., De Neve, W., Ro, Y.M.: Near-Duplicate Video Clip Detection Using Model-Free Semantic Concept Detection and Adaptive Semantic Distance Measurement. IEEE Transactions on Circuits and Systems for Video Technology 22(8), 1174–1187 (2012) ISSN 1051-8215

17. Hoi, C.-H., Wang, W., Lyu, M.R.: A Novel Scheme for Video Similarity Detection. In: Bakker, E.M., Lew, M., Huang, T.S., Sebe, N., Zhou, X.S. (eds.) CIVR 2003. LNCS, vol. 2728, pp. 373–382. Springer, Heidelberg (2003)

18. Wu, X., Ngo, C.-W., Hauptmann, A.G., Tan, H.-K.: Real-Time Near-Duplicate Elimination for Web Video Search With Content and Context. IEEE Transactions on Multimedia 11(2), 196–207 (2009) ISSN 1520-9210

19. Chaudhary, V., Sureka, A.: Contextual Feature Based One-Class Classier Approach for Detecting Video Response Spam on YouTube. In: Eleventh Annual Conference on Privacy, Security and Trust, PST (2013)

20. Jansohn, C., Ulges, A., Breuel, T.M.: Detecting pornographic video content by combining image features with motion information. In: Proceedings of the 17th ACM International Conference on Multimedia, MM 2009, pp. 601–604. ACM, New York (2009)

21. Fu, T., Huang, C.-N., Chen, H.: Identification of extremist videos in online video sharing sites. In: IEEE International Conference on Intelligence and Security Informatics, ISI 2009, pp. 179–181 (2009)

22. Sureka, A., Kumaraguru, P., Goyal, A., Chhabra, S.: Mining YouTube to Discover Extremist Videos, Users and Hidden Communities. In: Proceedings 6th Asia Information Retrieval Societies Conference, AIRS 2010, Taipei, Taiwan, December 1-3, pp. 13–24 (2010)

23. Dadvar, M., Trieschnigg, D., Ordelman, R., de Jong, F.: Improving cyberbullying detection with user context. In: Serdyukov, P., Braslavski, P., Kuznetsov, S.O., Kamps, J., Rüger, S., Agichtein, E., Segalovich, I., Yilmaz, E. (eds.) ECIR 2013. LNCS, vol. 7814, pp. 693–696. Springer, Heidelberg (2013)
24. Educause Learning Initiatives, 7 things you should know about...YouTube (September 2006), http://www.educause.edu/library/resources/7-things-you-should-know-about-youtube
25. US copyright Office, The Digital Millennium Copyright Act of 1998, U.S. Copyright Office Summary (December 1998), http://www.copyright.gov/legislation/dmca.pdf

High Dimensional Big Data and Pattern Analysis:
A Tutorial

Choudur K. Lakshminarayan

HP Software R&D, USA
Choudur.Lakshminarayan@hp.com

Abstract. Sensors and actuators embedded in physical objects being linked through wired/wireless networks known as *"internet of things"* are churning out huge volumes of data (McKinsey Quarterly report, 2010). This phenomenon has led to the archiving of mammoth amounts of data from scientific simulations in the physical sciences and bioinformatics, to social media and a plethora of other areas. It is predicted that over 30 billion devices with 200 billion intermittent connections will be connected by 2020. The creation and archival of the massive amounts of data spawned a multitude of industries. Data management and up-stream analytics is aided by data compression and dimensionality reduction. This review paper will focus on some foundational methods of dimensionality reduction by examining in extensive detail some of the main algorithms, and points the reader to emerging next generation methods that seek to identify structure within high dimensional data not captured by 2nd order statistics.

Keywords: Multivariate Analysis, Dimensionality Reduction, Projections, Principal Component Analysis, Factor Analysis, Canonical correlation Analysis, Independent Component Analysis, Exploratory Projection Pursuit.

1   Introduction

Needless to say, "Big Data" is the next frontier in scientific exploration and advancement. A recent report by a National Academy of Sciences commissioned study examines the challenges and opportunities [1]. The range of problems includes back-end systems for the ingestion and storing of large volumes of structured, semi-structured, and unstructured data (static and time-aware) in various formats and sources, to query engines, and *analytics* operations in the front-end. The volume to be processed at the speed of business requires parallel computing by distributed processing, a common set of analytical methods for repeatable analyses and rewriting existing algorithms to adapt to scale.

As the number of variables which purportedly describe a phenomenon, as well as frequency of sampling keeps increasing, it has become a challenge to tease out that subset of variables which indeed capture the dynamics and structure of the underlying phenomenon. Towards that end, data reduction techniques have become the *mainstay* of statistical data pre-processing. So we provide a tutorial review of some of the

V. Bhatnagar and S. Srinivasa (Eds.): BDA 2013, LNCS 8302, pp. 68–85, 2013.

foundational methods in dimensionality reduction in detail and point the reader to the next generation of algorithms. The field of dimensionality reduction is vast, and so we limit the scope of the paper to popular dimensionality reduction techniques such as *principal component analysis* (PCA), *Factor Analysis* (FA), *Canonical Correlation Analysis* (CCA) *Independent Component Analysis* (ICA), and *Exploratory projection Pursuit* (PP). We have chosen these methods because the vast majority of practitioners utilize them in daily applications. In this tutorial, we will study PCA, FA, CCA, ICA, and PP and relationships among them in some detail.

Since dimension reduction is not only desirable, but paramount, how should we go about it? Perhaps, a simple approach is to find a lower-dimensional embedding in which the data truly resides, while eliminating extraneous variance. This can be achieved by projecting the data by linear transformations into lower dimensional subspaces by maximizing a suitable objective function. This *genre* of algorithms is known as *projective methods* [13]. In the transformed domain the data is more interpretable as non-informative sources of variation can be eliminated, while retaining principal directions of variance. The other approach to dimensionality reduction is to exploit polynomial moments to unravel the hidden structure in the data. Well known projective methods are based on the covariance (2nd order cross moments) which only capture the linear structure in the data. Methods that go beyond 2nd order moments are exploratory projection pursuit, independent component analysis, and principal curves and surfaces [7,8]. A class of methods known as *manifold learning*, extract low-dimensional structure from high dimensional data in an unsupervised manner. These techniques typically try to unfold the underlying manifold ($\mathcal{M}$) into a lower dimensional space so that Euclidean distance in the new space is a meaningful measure of distance between pairs of points [16]. These methods have implications in making clustering methods more effective in the transformed space. In the following sub-sections, we will briefly introduce the techniques presented in this paper. The contents of the paper assume that the reader is familiar with elementary linear algebra, elementary probability theory, mathematical statistics, and multivariate analysis.

1.1   Principal Component Analysis

Principal components analysis (PCA) is one member of a family of methods for dimensionality reduction. It is a technique that involves transformations of set of variables into a smaller set of uncorrelated variables, while retaining intrinsic information in the original data set by exploiting correlations among the variables [2,3,15]. PCA is merely a linear projection of a set of observed variables on to basis vectors which turn out to be Eigen vectors under the average mean square error (MSE) objective function. PCA is one of the simplest and most common ways of doing dimensionality reduction. It is also one of the oldest, and has been variously alluded to in many fields as the Karhunen-Loève transformation (KLT) in communications, the Hotelling transformation, and latent semantic indexing (LSI) in text mining. But the *moniker* principal component analysis is the most popular.

1.2   Factor Analysis

Factor Analysis (FA) is a technique to find relationships between a set of observed variables and set of *latent* factors. The Factor analytic model is based entirely on the covariance matrix of the observed variables like the PCA models we studied in an earlier section. The key idea behind factor analysis is that multiple observed variables have similar patterns of responses because of their association with an underlying set of latent variables; the factors, which cannot be easily measured. For example, responses to questions about occupation, education, and income, are all associated with the latent variable socioeconomic stratum. In a factor analysis model, the number of factors always equal to the number of variables. Each factor contributes to a certain amount of the overall variance in the observed variables. The factors are then arranged in the decreasing order of variance explained. In a factor analytic model, each observed variable $\{X_i\}_{i=1}^p$ is expressed as a sum of latent factors $\{F_i\}_{i=1}^p$, known as *common* factors, and an error term $\{\epsilon_i\}_{i=1}^p$, known as *specific* variance. The specific variance accounts for the unexplained variance in the observed variable. Mathematically, the observed variables are projected onto a set of basis vectors $\{\ell_{ij}\}_{i,j=1}^p$ known as *loadings* in the FA literature. Under some assumptions on the latent factors, the loadings are the Eigen vectors obtained by decomposing the covariance matrix Σ. Typically spectral decomposition [2] is applied to Σ to obtain Eigen value, Eigen vector pairs $\{\lambda_i, \vec{l}_i\}_{i=1}^p$. The Eigen value is a measure of how much of the variance in the observed variables a factor explains. Any factor with an Eigen value ≥ 1 explains more variance than any single observed variable. In the exploratory mode, FA can be used to subset similar variables by examining the factor loadings on the original observed variables.

1.3   Canonical Correlation Analysis (CCA)

Canonical correlation analysis (CCA) proposed by Professor Harold Hotelling in 1936 is a method for exploring linear relationships between two sets of multivariate variables (vectors), measured on the same object/entity. It finds two bases, one for each variable set such that the correlation between the inner products (linear projections) between the two bases and the two variable sets is maximized. The dimensionality of these new bases is less than or equal to the smallest dimensionality of the two variables. Succinctly, CCA reduces pairs of high dimensional variables into a smaller set of linear combinations which are more amenable for interpretation.

1.4   Independent Component Analysis (ICA)

All the methods introduced are based on 2nd order statistics (correlation structure). Independent component analysis (ICA) [9] in contrast attempts to reduce dependencies among higher order moments thereby increasing statistical independence among the original variables. It is a technique for identifying an underlying set of hidden factors from a multivariate set of random variables. It is among one of the popular techniques used in blind source separation [9]. An observed set of observations are assumed to be linear mixtures of hidden latent factors, and the mixing coefficients are unknown. ICA departs from previously known projective methods in that it assumes that it exploits higher order moments beyond the 2nd order for identifying unknown factors (components) of the hidden mixture, besides assuming non-Gaussian distributions generating the data. In one version of the ICA, the latent sources are assumed to be non-Gaussian and independent. The objective function to estimate the unknown coefficients is parametrized by a likelihood function. The non-linear log-likelihood is then used to estimate the unknown coefficients by any of the stochastic gradient methods [11].

1.5   Projection Pursuit (PP)

Projection pursuit seeks to identify hidden structure in high dimensional data by using projections in lower dimensions that capture interesting features. The *interestingness* is determined by a numerical index known as the projection index. Techniques such as PCA, FA, and CCA depend on rotation, and scaling, to obtain linear projections. If the data vector $X \in \mathbb{R}^p$ observes a certain probability law, their sum $\vec{x}'\vec{w}$ would follow a Gaussian distribution by the central limit theorem [5]. And it is well known that the Gaussian is fully specified when the first two moments (mean, and covariance) are known. So, these methods capture only the linear structure in the data. Projection pursuit seeks to unravel the non-linear hidden structure by leveraging polynomial moments, and it is in this sense that PP departs from other projective methods.

2   Projective Methods and Dimensionality Reduction

2.1   Principal Component Analysis

Principal Component Analysis involves linear combinations of the p features x_1, x_2,....,x_p of an input pattern vector that are mean-centered. Geometrically, the linear combinations are obtained by rotating the original system with features x_1, x_2,....,x_p as the coordinate axes, thereby resulting in a new rotated coordinate system. The axes of the rotated coordinate system represent the directions with maximum variability. This lets elimination of low-variability coordinate axes to reduce the dimensionality of the original data. Although p principal components are required to account for the total system variability, majority of the variation is captured by a smaller number m. The m principal components can then replace the original p features. Thus, the original data set consisting of p features with n measurements each is replaced by p principal components with n measurements each. Thus principal components are vectors that span a lower m dimensional subspace. Material for this section has been adapted from [2,3,6,15] and the reader is encouraged to refer to these references.

Consider the random vector $X = (x_1, x_2, \cdots, x_p)$ with covariance matrix Σ whose Eigen values are $\{\lambda_i\}_{i=1}^p$, where each λ_i is ≥ 0. Let $\{z_i\}_{i=1}^p$, be a set of vectors obtained by composing linear combinations of the original features. Mathematically they are given as:

$$z_1 = w_{11}x_{11} + w_{21}x_{12} + \cdots + w_{p1}x_{1p}, w_{11}x_{21} + w_{21}x_{22} + \cdots + w_{p1}x_{2p}, \cdots w_{11}x_{n1} + w_{21}x_{n2} + \cdots + w_{p1}x_{np}$$

$$z_2 = w_{12}x_{11} + w_{22}x_{12} + \cdots + w_{p2}x_{1p}, w_{12}x_{21} + w_{22}x_{22} + \cdots + w_{p2}x_{2p}, \cdots, w_{12}x_{n1} + w_{22}x_{n2} + \cdots + w_{p2}x_{np}$$

$$\vdots$$

$$z_p = w_{1p}x_{11} + w_{2p}x_{12} + \cdots + w_{pp}x_{1p}, w_{1p}x_{21} + w_{2p}x_{22} + \cdots + w_{pp}x_{2p}, \cdots w_{1p}x_{n1} + w_{2p}x_{n2} + \cdots + w_{pp}x_{np}$$

Where $z_i = (z_{i1}, z_{i2}, \cdots, z_{ip})$ is the ith linear combination.

What we notice above is that each feature vector, say $\vec{x}_1 = (x_{11}, x_{12}, \cdots, x_{1p})$ is projected onto a vector $\vec{w} = (w_{11}, w_{21}, \cdots, w_{p1})$ given by; $\vec{x}_1'\vec{w}$ which is a simple inner product. The vector $\vec{w}$ is such that $\vec{w}'w = 1$. It is clear that the mean of the vectors $\{z_i\}_{i=1}^p$ is 0 since the x's are mean-centered. Consider for example, the vector z_i. The mean $\bar{z} = \frac{\sum_{j=1}^p \sum_{i=1}^n w_{j1}x_{ij}}{n} = \sum_{j=1}^p w_{j1}\frac{\sum_{i=1}^n x_{ij}}{n} = 0$ as the x's are mean-centered.

In matrix form, the linear combinations $\{z_i\}_{i=1}^p$ can be written as $Z = [XW]'$

The variance of Z can be expressed in matrix form as:

$$\sigma_z^2 = \frac{1}{n}[XW]'[XW] \xrightarrow{yields} W'\frac{X'X}{n}W \qquad (1)$$

To derive principal components from linear combinations (projections), we invoke the notion of average *mean square error* (MSE). That is; we are searching for those projections that have the smallest mean square distance between the original feature vectors and their projections. Mathematically, we want to choose the vector $\vec{w}$ such that the variance σ_z^2 is minimized. To find the $\vec{w}$ that maximizes the variance (σ_z^2), we utilize constrained optimization by Lagrange multipliers [17]. Maximize σ_z^2 subject to $\vec{w}'\vec{w} = 1$

$$\mathcal{L}(\vec{w}, \lambda) = \sigma_z^2 - \lambda(\vec{w}'\vec{w} - 1) \qquad (2)$$

$$\frac{\partial \mathcal{L}(\vec{w}, \lambda)}{\partial \vec{w}} = 2\frac{X'X}{n}\vec{w} - 2\lambda\vec{w} \qquad (3)$$

$$\frac{\partial \mathcal{L}(\vec{w}, \lambda)}{\partial \lambda} = \vec{w}'\vec{w} - 1 \qquad (4)$$

Let $\frac{X'X}{n} = S$ and setting $\frac{\partial \mathcal{L}(\vec{w},\lambda)}{\partial \vec{w}} = 0$ implies $S\vec{w} = \lambda\vec{w}$, the characteristic equation that links Eigen values and Eigen vectors. Therefore, the desired vector $(\vec{w})$ is the Eigen vector of the covariance matrix (S). These maximizing Eigen vectors will be associated with the largest Eigen values(λ). Since S is a covariance matrix, it is symmetric and positive definite. A matrix is said to be positive definite, if $\vec{x}'S\vec{x} > 0$ for any $\vec{x}$. It is well known that a symmetric, positive definite matrix has positive Eigen values and the corresponding Eigen vectors are orthogonal. The first principal component is the axis along which the data has the most variance, and corresponds to a projection on the Eigen vector with the largest Eigen value. Similarly, the 2nd principal component is the axis with the 2nd largest variance and is associated with the with the Eigen vector with 2nd largest Eigen value, and so on. And we obtain p principal components as the covariance matrix is of order (pxp). Since the Eigen vectors are orthogonal, the projections (principal components) are all uncorrelated with each other. As each principal component captures proportion of variance in the data along its axis; those components corresponding to low variance may be dropped. Thus a set of $q \ll p$ fewer components may be chosen, resulting in dimensionality reduction. In a practical setting, the Eigen values obtained by solving, $S\vec{w} = \lambda\vec{w}$, are given by; $\{\lambda_i\}_{i=1}^p$ are ordered. The ordered set of Eigen values, from the smallest to the largest are$(\lambda_{(1)}, \lambda_{(2)}, \cdots, \lambda_{(p)})$. The variance explained by each successive principal component is obtained by calculating the ratio;

$$\psi = \frac{\sum_{j=1}^i \lambda_j}{\sum_{j=1}^p \lambda_j}, j = 1,2,\cdots,p. \tag{5}$$

When ψ exceeds 0.8 (say), then the number of principal components is equal to i for which 0.8 is attained. The number 0.8 corresponds to 80% of variance in the data. The experimenter is at liberty to choose the cut-off value appropriate for the application$(x\%)$. Many times, a graph known as the *scree* plot is drawn to select the appropriate number of principal components. The scree plot is merely a graph, where the X-axis represents the numbers $(1,2,\cdots,p)$-which are the indexes of the Eigen values and the Y-axis is the cumulative variance(ψ). The number of principal components is chosen by locating the elbow in the curve beyond which the additional variance is negligible. Fig. 1 is an illustration of a scree plot. The vertical arrow marks the point when the cumulative variance stabilizes (flattens).

In conclusion, while PCA is a useful data reduction technique, care should be exercised in extracting meaning out of the components, which are simply linear projections of the original data. If the end goal is to classify high dimensional objects/entities to one of a several classes (as in a classification problem), using PCA for data reduction is *fair game* and perhaps required.

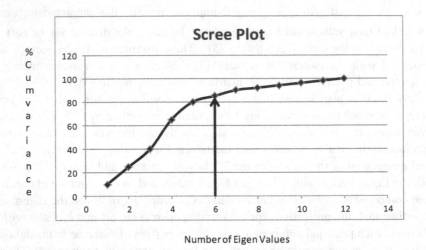

Number of Eigen Values

Fig. 1. Graph of Cumulative Variance versus Number of Eigen Values

2.2   Canonical Component Analysis

Canonical Correlation Analysis (CCA) proposed by Professor Harold Hotelling in 1936 is a method to correlate two different set of variables by projective transformations [4, 14]. It seeks to reduce pairs of high dimensional vectors into a few pairs of highly-correlated linear combinations of vectors known as *canonical* variables. Thus CCA can be construed as a feature reduction technique, while its origins was in being able to find relationships between manifestly different sets of variables, such as those related to government policies and economic impact. Operationally, CCA involves projecting each set of multi-dimensional variables $(\vec{x}, \vec{y})$ onto basis vectors $(\vec{w}_x, \vec{w}_y)$ such that the correlation measure (ρ) between the projected vectors is maximized. The projections are given by $\vec{x}'\vec{w}_x$ and $\vec{y}'\vec{w}_y$. The idea is to find pairs of linear projections that are maximally correlated. The next iteration, we find those projections that are maximally correlated, but uncorrelated with the first pair, and the procedure continues until we find correlated projections that are uncorrelated with the predecessor pairs. Fig. 2 is a pictorial representation of Canonical Correlation Analysis. Two

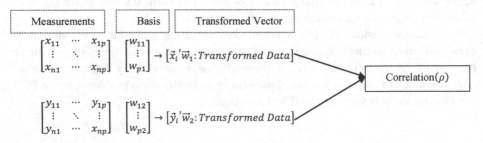

Fig. 2. A Pictorial representation of Canonical Correlation Analysis

sets of variables (X, Y) are projected on to two basis vectors (W_x, W_y) to yield linear projections, $(X'W_x, Y'W_y)$. We then determine the optimal values of the basis vectors; (W_x, W_y) that maximizes the correlation (ρ) between the projections.

Operationally, we are seeking vectors $\vec{w}_x$ and $\vec{w}_y$ such that:

$$\rho = \max_{\vec{w}_x, \vec{w}_y} \frac{E(\vec{x}'\vec{w}_x \, \vec{y}'\vec{w}_y)}{\sqrt{E(\vec{x}'\vec{w}_x)^2 E(\vec{y}'\vec{w}_y)^2}} \tag{6}$$

The above expression can be further simplified for simplicity of expression by simple algebraic manipulations.

$$\rho = \max_{\vec{w}_x, \vec{w}_y} \frac{E\left(\vec{w}_x'\vec{x}\,\vec{y}'\vec{w}_y\right)}{\sqrt{E\left(\vec{w}_x'\vec{x}\vec{x}'\vec{w}_x\right) E\left(\vec{y}'\vec{w}_y y'\vec{w}_y\right)}} \xrightarrow{yields} \max_{\vec{w}_x, \vec{w}_y} \frac{\vec{w}_x' E(\vec{x}\,\vec{y}')\vec{w}_y}{\sqrt{\vec{w}_x' E(\vec{x}\vec{x}')\vec{w}_x E\left(\vec{w}_y' E(\vec{y}\vec{y}')\vec{w}_y\right)}} \tag{7}$$

$$\rho = \max_{\vec{w}_x, \vec{w}_y} \frac{\vec{w}_x' \Sigma_{xy} \vec{w}_y}{\sqrt{\vec{w}_x' \Sigma_{xx} \vec{w}_x \vec{w}_y' \Sigma_{yy} \vec{w}_y}} \tag{8}$$

where $\Sigma_{xx}, \Sigma_{yy}, \Sigma_{xy}$ are respectively, the variances and covariance between the random variables .

The maximum of ρ is the canonical correlation obtained by maximizing over $(\vec{w}_x, \vec{w}_y)$. We note that the canonical correlation is invariant to scaling the basis by a constant (c). This can easily seen by re-scaling to $c\vec{w}_x, c\vec{w}_y$ and substituting in (8). Thus we maximize the canonical correlation subject to the constraints $\vec{w}_x' \Sigma_{xx} \vec{w}_x = 1$ and $\vec{y}'\vec{w}_y y'\vec{w}_y = 1$. Since we are seeking an optimization solution under constraints above, the Lagrangian formulation is as follows:

$$\mathcal{L}(w_x, w_y, \lambda) = \vec{w}_x' \Sigma_{xy} \vec{w}_y - \frac{\lambda_x}{2}\left(\vec{w}_x' \Sigma_{xx} \vec{w}_x - 1\right) - \frac{\lambda_y}{2}\left(\vec{w}_y' \Sigma_{yy} \vec{w}_y - 1\right) \tag{9}$$

Finding the derivatives of $\mathcal{L}(\cdot)$ with respect to $\vec{w}_x, \vec{w}_y$ and setting them equal to zero yields;

$$\frac{\partial \mathcal{L}(w_x, w_y, \lambda_x, \lambda_y)}{\partial w_x} = \Sigma_{xy}\vec{w}_y - \lambda_x \Sigma_{xx}\vec{w}_x = 0 \tag{10}$$

$$\frac{\partial \mathcal{L}(w_x, w_y, \lambda_x, \lambda_y)}{\partial w_y} = \Sigma_{xy}\vec{w}_x - \lambda_y \Sigma_{yy}\vec{w}_y = 0 \tag{11}$$

To solve this system of linear equations, we multiply, (10) by $\vec{w}_x'$ and the (11) by $\vec{w}_y'$ yielding,

$$\vec{w}_x' \Sigma_{xy}\vec{w}_y - \lambda_x \vec{w}_x' \Sigma_{xx}\vec{w}_x = 0 \tag{12}$$

$$\vec{w}_y'\Sigma_{xy}\vec{w}_x - \lambda_y\vec{w}_y'\Sigma_{yy}\vec{w}_y = 0 \tag{13}$$

Subtracting, the (13) from the (12), gives:

$$(\lambda_y - \lambda_x)(\vec{w}_y'\Sigma_{yy}\vec{w}_y - \vec{w}_x'\Sigma_{yy}\vec{w}_x) = 0 \tag{14}$$

Applying the constraints, $\vec{w}_x'\Sigma_{xx}\vec{w}_x = 1$ and $\vec{w}_y'\Sigma_{xx}\vec{w}_y = 1$, we obtain, $\lambda_y = \lambda_x$.

Also from (10) we get $\vec{w}_x = \frac{\Sigma_{xx}^{-1}\Sigma_{xy}\vec{w}_y}{\lambda_x}$. Substituting this value of $\vec{w}_x$ in in (11) we have:

$$\frac{\Sigma_{xy}'\Sigma_{xx}^{-1}\Sigma_{xy}}{\lambda_x}\vec{w}_y = \lambda_y\Sigma_{yy}\vec{w}_y \xrightarrow{yields}$$

$$(\Sigma_{xy}'\Sigma_{xx}^{-1}\Sigma_{xy})\vec{w}_y = \lambda_x\lambda_y\Sigma_{yy}\vec{w}_y \xrightarrow{yields} \lambda^2\Sigma_{yy}\vec{w}_y \tag{15}$$

since $\lambda_y = \lambda_x = \lambda$. Note that I use Σ_{xy}' instead of Σ_{xy} in (15) as it is a symmetric matrix.

$(\Sigma_{xy}'\Sigma_{xx}^{-1}\Sigma_{xy})\vec{w}_y = \lambda^2\Sigma_{yy}\vec{w}_y$ in (15) is reminiscent of an Eigen equation. It is known as a generalized Eigen equation. This can be reduced to the form $Ay = \lambda y$, by noting that the matrix Σ_{yy} is symmetric and positive definite and can be expressed as the product $L_{yy}L_{yy}'$ (Cholesky Decomposition) [11]. Also, let $\vec{u}_y = L_{yy}'\vec{w}_y$ and re-writing (15), we have:

$$\Sigma_{xy}'\Sigma_{xx}^{-1}\Sigma_{xy}(L_{yy}^{-1})'\vec{u}_y = \lambda^2 L_{yy}'\vec{w}_y = \lambda^2\vec{u}_y \tag{16}$$

Clearly, this of the form $Ay = \lambda y$ is the Eigen equation seen in standard linear algebra! The Eigen equation can be used to find the $(\vec{w}_y, \vec{w}_x)$ to find the co-ordinate system that optimizes the correlation between the linear combinations of the two sets of vectors $(\vec{x}, \vec{y})$. To apply the theory developed above in a practical setting, the unknown population quantities, Σ_{xx}, Σ_{yy}, and Σ_{xy} are replaced by their sample counterparts, S_{xx}, S_{yy}, and S_{xy} respectively. Alternatively, one may use the correlation matrices, R_{xx}, R_{yy}, and R_{xy} as the roots/solutions derived from the application of the two representationss is the same.

In conclusion, CCA can be applied to large data sets where the correlations among the linear combinations of two sets of variables may reveal latent structures in the data that may not be captured by pair-wise correlations among the original variables. An added advantage of CCA is that it can identify relationships among observed variables and underlying latent factors/motivations. For example, it is applied in marketing analytics to understand the relationship between pricing (observed variable) and attributes such as form factor, ease of using, appeal, and other features (latent factors) attractive to a target segment of the market.

2.3   Factor Analysis

Consider a random vector X with p components with mean vector (μ) and covariance matrix(Σ). The factor analysis model is given by

$$
\begin{aligned}
X_1 - \mu_1 &= l_{11}F_1 + l_{12}F_2 + \cdots + l_{1m}F_m + \epsilon_1 \\
X_2 - \mu_2 &= l_{21}F_1 + l_{22}F_2 + \cdots + l_{2m}F_m + \epsilon_2 \\
&\vdots \\
X_2 - \mu_2 &= l_{21}F_1 + l_{22}F_2 + \cdots + l_{2m}F_m + \epsilon_2
\end{aligned} \tag{17}
$$

Where $\{X_i, \mu_i, F_i, \epsilon_i\}_{i=1}^p$ are the p observed variables, unknown means of the observed variables, the hidden latent factors called the common factors and $\{\epsilon_i\}_{i=1}^p$ are the specific variances respectively. And $\{l_{ij}\}_{i,j}^m$ are the factor loadings. Fig. 3. is a graphical illustration of the FA model, where subsets of the observed variables are captured by the factors, and the specific variances (ϵ_i) are associated with the individual variables (X_i). Clearly from the set of equations (17), the observed variable resolves into a factor component(F_i) and a specific variance component (error) shown in Fig. 4.

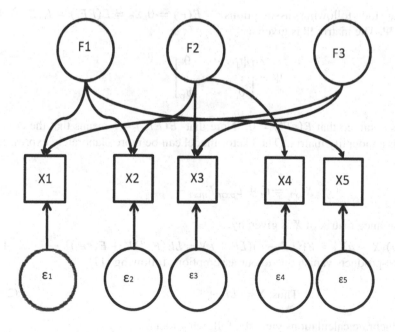

Fig. 3. A graphical representation of the factor analysis model

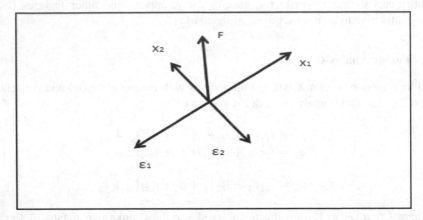

Fig. 4. Resolution of an observation vector (x), into a common factor (F) and error components $\varepsilon_1, \varepsilon_2$

The FA model postulates that a vector X is decomposable into a set of common factors $F_i, (i = 1,2, \cdots, m)$ and specific factors $\{\epsilon_i\}_{i=1}^{p}$. In matrix terms, the FA model in (17) can be written as;

$$X - \mu = LF + \varepsilon \qquad (18)$$

We make the following assumptions. $E(F) = 0, \Sigma_F = E(FF') = I, , E(\varepsilon) = 0, E(\varepsilon\varepsilon') = \Psi$. The matrix Ψ is given as:

$$\Psi = \begin{bmatrix} \psi_1 & \cdots & 0 \\ \vdots & \ddots & \vdots \\ 0 & \cdots & \psi_p \end{bmatrix}$$

Also, it is assumed that $E(\varepsilon F') = 0$. Note that $E(FF') = I$ means that the covariance of F is an identity matrix. The Factor model can be more elaborately expressed as:

$$X_{px1} = \mu_i + L_{pxm}F_{mx1} + \varepsilon_{px1} \qquad (19)$$

The covariance matrix of X is given by;

$E\{(X - \mu)(X - \mu)'\} = E\{(LF + \varepsilon)(LF + \varepsilon)' = LE(FF')L' + E(\varepsilon\varepsilon')\} \xrightarrow{yields} LL' + \Psi$. The cross-products vanish due to our assumptions following (17)

$$\text{Thus, } \Sigma = LL' + \Psi \qquad (20)$$

Simple algebraic calculations yield the following identities;

$$Var(X_i) = l_{i1}^2 + l_{i2}^2 + \cdots + l_{im}^2 + \psi_i \qquad (21)$$

and

$$cov(X_i, X_k) = l_{i1}l_{k1} + l_{i2}l_{k2} + \cdots + l_{im}l_{km} \tag{22}$$

and $cov(X_i, F_j) = l_{ij}$.

It is clear from the model formulation that factor analysis attempts to reproduce the $p + p(p+1)/2$ variances and covariance using pm factor loadings and specific variances. So the choice of the number of factors (m) is mighty important.

Heretofore, our discussion focused on what is known as the population model in the statistics literature. Since $\Sigma = LL' + \Psi$ is unknown, it is estimated by the sample covariance (S). We use the sample covariance matrix (S) to estimate the factor loadings. The loadings estimated from the sample are called sample loadings. And the specific variances may be construed as the unexplained sample variance.

$$S = \begin{bmatrix} l_{11} & \cdots & l_{1m} \\ \vdots & \ddots & \vdots \\ l_{p1} & \cdots & l_{pm} \end{bmatrix} \begin{bmatrix} l_{11} & \cdots & l_{p1} \\ \vdots & \ddots & \vdots \\ l_{1m} & \cdots & l_{pm} \end{bmatrix} + \begin{bmatrix} \psi_1 & \cdots & 0 \\ \vdots & \ddots & \vdots \\ 0 & \cdots & \psi_p \end{bmatrix} \tag{23}$$

Therefore,

$$S_{ii} = \underbrace{\sum_{j=1}^{m} l_{ij}^2}_{communality} + \underbrace{\psi_i}_{specific\,variance} \tag{24}$$

Communality is the sum of the squared factor loadings on the m factors for a given variable. It is the variance in that variable accounted for, by the m factors. Another way to understand *communality* is that is a measure of percent of variance in a given variable explained by the m factors jointly and may be interpreted as the reliability of the indicator (latent factors).

In order to obtain a sample based solution, we use PCA. This is achieved by applying spectral decomposition [3] to the sample covariance matrix (S). The PCA approach decomposes S in terms of Eigen values and Eigen vector pairs. Mathematically, spectral decomposition of a pxp symmetric matrix is given as: $S_{pxp} = \lambda_1 e_{1(px1)} e'_{1(1xp)} + \lambda_2 e_{2(px1)} e'_{2(1xp)} + \cdots + \lambda_2 e_{p(px1)} e'_{p(1xp)}$, where λ_i, ($i = 1, 2, \cdots, p$) are the Eigen values and e_i is the ith Eigen vector. This representation of the sample covariance matrix is known as the famous spectral decomposition. For the FA model to be useful, only the top $m \ll p$ Eigen vectors are retained, and the specific variances (ψ_i) are assumed to be negligible. In some cases the specific variances are assumed to be non-negligible as well. Furthermore, the spectral decomposition of S can be written as:

$$S_{pxp} = \left[\sqrt{\lambda_1} e_1 \mid \sqrt{\lambda_2} e_2 \cdots \mid \sqrt{\lambda_p} e_p\right] \begin{bmatrix} \sqrt{\lambda_1} e_1 \\ \vdots \\ \sqrt{\lambda_p} e_p \end{bmatrix} \tag{25}$$

Let us assume that the specific variances (ψ_i) are non-negligible. Then the FA model is given by

$$S_{pxp} = \left[\sqrt{\lambda_1}e_1 \mid \sqrt{\lambda_2}e_2 \cdots \mid \sqrt{\lambda_p}e_p\right] \begin{bmatrix} \sqrt{\lambda_1}e_1 \\ \vdots \\ \sqrt{\lambda_p}e_p \end{bmatrix} + \begin{bmatrix} \psi_1 & \cdots & 0 \\ \vdots & \ddots & 0 \\ 0 & 0 & \psi_p \end{bmatrix} \tag{26}$$

If we assume the 2nd matrix (Ψ) to be negligible,

$$S_{pxp} \cong \left[\sqrt{\lambda_1}e_1 \mid \sqrt{\lambda_2}e_2 \cdots \mid \sqrt{\lambda_p}e_p\right] \begin{bmatrix} \sqrt{\lambda_1}e_1 \\ \vdots \\ \sqrt{\lambda_p}e_p \end{bmatrix} = L'L \tag{27}$$

Examining the decomposition equation for the sample covariance matrix (S), the loadings l_{ij} is the solution to the equation, $S_{pxp} = L'L$. The loadings appearing as coefficients in the observed variables (X_i) are used to impute meaning to factors and also identify sub groupings of observed variables.

The FA model, while useful for identifying for sub-groupings of original variables, is beset with some ambiguities. The loadings, L are only unique up to rotation. Consider an orthogonal matrix (R) such that $R'R = RR' = I$. The matrix R is a rotation matrix [17]. We saw that the FA model $X - \mu = LF + \varepsilon$ yields the covariance matrix, $\Sigma = LL' + \Psi$. Which can be rewritten as:

$$\Sigma = L\,RR'L' + \Psi \xrightarrow{\text{yields}} L_R L_R' + \Psi \tag{28}$$

where $L_R = LR$. So, we notice that both L and L_R yield the same covariance matrix Σ. So a rotated version of L leads to a set of loadings with the same covariance leading to an obvious ambiguity. So, in applying FA models, if the initial loadings do not yield a satisfactory solution in identifying a reduced set variable sub-groupings or meaningful interpretations, the experimenter can apply rotations such that the loadings L can be redistributed among the factors to possibly obtain more meaningful results. This task is often accomplished by such procedures as *varimax*, and *promax* rotations. Commercial and open source tools such as SAS, MATLAB, and open source program R provide this feature. The material for this section has largely been adapted from [3]. The reader is encouraged to consult it for a detailed exposition.

2.4   Independent Component Analysis

Independent component analysis (ICA) is a versatile technique that can be used for data reduction. ICA is a tool for discovering underlying latent factors that are statistically independent and do not observe the Gaussian law of errors to paraphrase [9]. While PCA and FA depend on the covariance matrix(Σ), ICA seeks projective directions that are statistically independent based on the probability distribution of the data and its higher order moments. Graphically, the ICA model given in Fig. 5, depicts the latent factors linearly combining to produce an output X_j at the jth node with the edge weight equal to l_{ij} connecting the ith latent source(F_i). ICA determines the optimal weights $\hat{l}_{ij}$. The weights may be construed as the correlation between the latent factor (F_i) and the observed output(X_j). More formally, let, X be a $p \times n$ matrix consisting of n observed samples of a p-dimensional vector $\vec{x}_i$, F is also a $p \times n$ matrix of consisting of n samples of a p-dimensional latent source vector $\vec{f}_i$. And L is a $p \times p$ matrix of unknown weights to be determined. The ICA model in the matrix form is given below.

$$\begin{pmatrix} x_{11} & \cdots & x_{1n} \\ \vdots & \ddots & \vdots \\ x_{p1} & \cdots & x_{pn} \end{pmatrix} = \begin{pmatrix} l_{11} & \cdots & l_{1p} \\ \vdots & \ddots & \vdots \\ l_{p1} & \cdots & l_{pp} \end{pmatrix} \begin{pmatrix} f_{11} & \cdots & f_{1n} \\ \vdots & \ddots & \vdots \\ f_{p1} & \cdots & f_{pn} \end{pmatrix} \tag{29}$$

Concisely, it is given by; $X_{(p \times n)} = L_{(p \times p)} F_{(p \times n)}$. In this equation, X, L, and F are the observed data, the unknown weights, and the unknown latent factors respectively. The objective is to estimate the unknown weights and factors optimally. Notice that unlike the FA model, the ICA does not explicitly consider the specific variances (ϵ). In other words, we are trying to seek, $F = WX$, where $W = L^{-1}$. So we can recover the latent sources(F_i), where $F_i = WX_i$ [18].

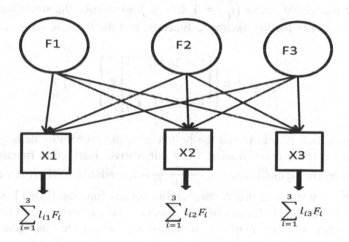

Fig. 5. Graphical illustration of an ICA model. The latent sources F_i are linearly combined to produce an output $\sum_{i=1}^{3} l_{ij} F_i$, where l_{ij} is the weight connecting the ith latent factor F_i to the ith output X_i.

Let the probability distribution of each source F_i be $g_F(\cdot)$. The joint probability distribution of the p independent latent sources and n independent samples (28) is

$$g(F) = \prod_{i=1}^{n} \prod_{j=1}^{p} g(f_{ij}) = \prod_{i=1}^{n} \prod_{j=1}^{p} g\left(\vec{l}_j' \vec{x}_i\right) |L^{-1}| \tag{30}$$

The above joint probability density of the latent sources in (30) is based on a well-known theorem on transformation of random variables from mathematical statistics [5]. The quantity $|L^{-1}|$ is known as the Jacobian of transformation which is the 2nd term in (30). It corresponds to the 2nd term in (31) of theorem 2.4.1. Note that the vector $(\vec{x}_i)$ is the ith observation and $\vec{l}_j$ is the corresponding weight vector in (29).

Theorem 2.4.1 If a random variable $X \sim f_X(x)$, then the transformed random variable,

$$Y = h(X) \sim f_X(Y) \left| \frac{\partial h^{-1}(y)}{\partial y} \right| \tag{31}$$

Given the observed data $\vec{x}_i, (i = 1, 2, \cdots, n)$, the log-likelihood function (ℓ) relative to the joint density $g(F)$ is denoted and written as:

$$\ell(l) = \sum_{i=1}^{n} \sum_{j=1}^{p} \log\left[g'\left(\vec{l}_j' \vec{x}_i \right) \right] + \log|L^{-1}| \tag{32}$$

You will notice that the log-likelihood function is written assuming that the observed samples, and therefore the unknown latent sources are independent. In order to determine the weight vector $(\vec{l}_j, j = 1, 2, \cdots, p)$, we invoke the stochastic gradient methods to maximize the log-likelihood function, and the iterative sequence is given by;

$$L \leftarrow L + \alpha \left\{ \begin{bmatrix} 1 - 2g\left(\vec{l}_1' \vec{x}_i \right) \\ \vdots \\ 1 - 2g\left(\vec{l}_p' \vec{x}_i \right) \end{bmatrix} \vec{x}_i \right\} + (L)^{-1} \tag{33}$$

In the application ICA to obtain the best results, the probability density function $g(\cdot)$ is assumed to be non-Gaussian. The cumulative distribution function (CDF) parametrized by the sigmoid function, $\frac{1}{1-exp(-f)}$ is a candidate CDF. It is well known from mathematical statistics that the probability density function (PDF) is simply the derivative of the CDF [5]. It can easily be checked that the derivative of the sigmoid does not result in the PDF of a Gaussian, which in its general form is; $\frac{1}{\sqrt{2\pi}} exp^{-\frac{1}{2\sigma^2}(f-\mu)^2}$, where (μ, σ^2) are the mean and variance respectively. Equation (33) is derived based on the assumption of the sigmoid function.

In the updating equation above, the parameter (α) is the learning rate. The 2nd term, $(L)^{-1}$ is obtained by finding the derivative of $\log|L^{-1}|$. On finding the optimal values of L, the latent factors can be constructed from $F_i = WX_i$, where $W = L^{-1}$. It is noted *en passant* that another typical application of ICA is for identifying latent sources (these correspond to mixed signals of voice samples captured by microphones placed in room-the famous cocktail party problem), our perspective and purpose here is different. We assume that there is a model consisting of a $p \times n$ matrix (X) of observed data which is a linear combination of latent sources. The idea is to identify a smaller set containing linear combinations of latent sources which explain the variance in the observed data.

In deriving the ICA model, we used a sigmoid function parametrization. However, if the application suggested a certain parametric form for the PDF, one should incorporate it into the joint density function, and derive the updating equations accordingly.

2.5   Projection Pursuit

Pursuant to our effort to discover hidden structure not captured by the covariance matrix(Σ), we introduce a popular technique called projection pursuit (PP) which is somewhat computationally intensive. Projection pursuit is a technique to reduce high-dimensional data by projecting on to a lower-dimensional space to reveal latent (hidden) structure in the higher dimensions [7]. Projection pursuit was first invented by Krushkal to discover *interesting* lower-dimensional projections. The notion of "interestingness" is parametrized by an index given by $I(w)$. If we recall, PCA, the goal there was to find axes of an ellipsoid (assuming the data is Gaussian) that corresponded to largest variation. So, the index $I(w)$ is the projection ($\vec{x}'\vec{w}$) of the data vector on to an Eigen vector($\vec{w}$) subject to $\vec{w}'\vec{w} = 1$. The "interestingness" in the data is of course the linear projections which are the principal axes parametrized by the Eigen vectors. For comprehensive detail and a beautiful exposition of PP, the reader is referred to [7]. We alluded to *hidden structure* in high dimensions. The Gaussian distribution being rotationally symmetric does not produce interesting projections. Because a linear projection ($\vec{w}'X$) where the random vector $X \in \mathbb{R}^p$, being a sum of random variables will again observe the Gaussian law by the central limit theorem. Therefore, a preponderance of linear projections do not reveal structure beyond the 2nd order moments. The projection index we seek is based on polynomial moments. The idea is to transform a projection($\vec{w}'X$) to $P = 2\Phi(\vec{w}'X) - 1$ where$\Phi(\cdot)$ denotes a Gaussian CDF. The transformation results in a Uniform distribution. The transformed projection is then compared against a Uniform random variable(U). A departure from the uniform distribution measured by $f_P(p) - g_U(u)$ is an indication of non-Gaussian structure. The symbols(f, g) are the distributions of the two random variables, (P, U)respectively. Operationally, it is the integral squared distance between the densities of P, U that is calculated. The integral square error statistic serves as a projection index. The reader is again encouraged to refer to [7].

3   Applications

In this section, we picked PCA to illustrate the importance of dimensionality reduction in a semiconductor manufacturing application. Signature analysis (SA) in semiconductor manufacturing is a statistical pattern recognition program designed to assign failed parts to one of several pre-determined root cause categories [10]. Engineers invest lots of time tracing back-end electrical parameter test failures to probable on-line root causes. It is desired to have an automated program based on sound statistical theory that enables the classification of a failing signature to a root cause category such that the probability of misclassification is minimized. Linear discriminant analysis (LDA) is an established parametric procedure that minimizes the probability of misclassification and allows the failure analysis engineer to state "The probability that a failing chip with a specific signature belongs to the kth root cause category is p %." But prior to applying LDA, a database of signatures is created. A signature is merely a feature vector of measurements obtained from a chip. Associated with each signature is a *label* which indexes the failing chip with an associated root cause. In many semiconductor manufacturing settings, the size of the signature vector is in excess of 400 features due to the number of tests conducted to ensure the reliability of the finished product. A majority of these tests are electrical measurements that are correlated to one another. So applying of PCA not only reduces the dimensionality of the signature vector, but also eliminates the collinearity (correlations) among the features since the principal components are orthogonal to one another. In the example below, chips are manufactured using the LinBiCMOS technology. LinBiCMOS is a CMOS technology with bipolar components (see Wikipedia for details about the semiconductor technologies). The chips were tested at 5 locations on a wafer. A wafer is an array of chips laid out as a matrix on a circular disc. The wafers are processed in batches of 20 are known as *lots*. The five locations known as test structures are at the top, center, bottom, left, and right (T,C,B,L,R) locations on the wafer. The electrical test measurements were approximately 125. Many of the measured features were correlated and thus redundant. We applied PCA to reduce the feature set to 34 principal components, which is a reduction of ~75%. An example using LDA to determine root cause of failures is shown in Table 1.

Table 1. Classification by Linear Discriminant Analysis

Lot Number	9745158
Device	XXXXXXXXX
technology	YY
Number of Wafers	24
Number of Sites	5
Number of Parameters	123

LDA by Site				
Wafer Number	Site Number	MD	Root Cause	probability
17	2	9.65	Missed N+S/D implant	0.980000
17	2	2.03.84	Missed Nwell Implant	0.000000
17	2	367.48	High Epi Doping	0.000000
17	2	408.77	Sidewall Overetch	0.000000

Two lots failing due to missing N+S/D implant were submitted to the automated signature analysis program for root cause identification (see the column headed, "probability" in Table 1). A signature of length 34 is applied to the program for pattern classification. The signature is from a certain device XXXXXXXXX belonging to technology YY. Table 1 shows the results of this analysis. The number of electrical test parameters measured for this technology is 123, but a signature of dimension 34 is applied for classification using PCA. Wafer 17 which failed some tests was applied to the SA program. The measurements were obtained from site 2 which corresponds to one of the locations (T,C,B,L,R).   Clearly, LDA identified the correct root cause, and the dimensionality reduction by PCA captured sufficient information to draw the correct inference!

References

1. Committee on the Analysis of Massive Data, Frontiers in Massive Data Analysis. National Academies Press (2013)
2. Shalizi, C.R.: Advanced Data Analysis from an Elementary Point of View (2013), http://www.stat.cmu.edu/~cshalizi
3. Johnson, R.A., Wichern, D.W.: Applied Multivariate Statistical Analysis, 3rd edn. Prentice Hall, Englewood Cliffs (1992)
4. Hotelling, H.: Relations Between Two Sets of Variables. Biometrika 28, 321–377 (1936)
5. Mood, A.M., Graybill, F.A., Boes, D.C.: Introduction to the Theory of Statistics, 3rd edn. McGraw-Hill (1974)
6. Hardle, W.K., Simar, L.: Applied Multivariate Statistical Analysis, 3rd edn. Springer (2011)
7. Friedman, J.H.: Exploratory Projection Pursuit. Journal of the American Statistical Association 82(397), 249–266 (1987)
8. Hastie, T., Tibshirani, R., Friedman, J.: The Elements of Statistical Learning, Data Mining, Inference, and Prediction. Springer (2001)
9. Hyvärinen, A., Karhunen, J., Oja, E.: Independent Component Analysis. Wiley, Inter-Science (2001)
10. Lakshminarayan, C.K., Baron, M.I.: Pattern Recognition in Large-Scale Data Sets: Application in Integrated Circuit Manufacturing. In: Bhatnagar, V. (ed.) BDA 2013. Springer, Heidelberg (2013)
11. Press, W.H., Flannery, B.P., Teukolsky, S.A., Vetterling, W.T.: Numerical Recipes in C, The Art of Scientific Computing. Cambridge University Press (1990)
12. Strang, G.: Linear Algebra and its Applications, 4th edn. Brooks/Cole Publishing Company (2005)
13. Burgess, C.J.C.: Dimension Reduction: A guided Tour. Foundation and Trends in Machine Learning 2(4), 275–365 (2010)
14. Hardoon, D.R., Szedmak, S., Shawe-Taylor, J.: Canonical correlation analysis; An overview with application to learning methods, technical report, CSD-TR-03-02, Dept. of Computer Science, Royal Holloway, University of London (2003)
15. Timm, N.H.: Applied Multivariate Analysis. Springer (2002)
16. Lee, J.A., Verleysen, M.: Nonlinear Dimensionality Reduction. Springer (2007)
17. Strang, G.: Introduction to Applied Mathematics. Wellesley-Cambridge Press (1986)
18. Ng, A.: Independent Component Analysis, CS229, Lecture Notes. Stanford University

The Role of Incentive-Based Crowd-Driven Data Collection in Big Data Analytics: A Perspective

Anirban Mondal

Xerox Research Center, India
anirban.mondal@xerox.com

Abstract. Big data analytics for effective decision-making entails significant amounts of data collection. Existing sensor-based data collection mechanisms are expensive to deploy due to the high initial fixed costs of installing large-scale sensor-based systems. Sensors also require maintenance, thereby further adding to the costs. Moreover, sensors cannot be cost-effectively installed at all possible locations. Furthermore, some data collection scenarios require human judgment, which sensors are not capable of providing. To address the limitations associated with sensor-based data collection mechanisms, this paper discusses the role of *incentive-based crowd-driven data collection* in big data analytics. Given the increasing prevalence and popularity of mobile devices coupled with the fact that mobile devices often come equipped with various kinds of sensors, crowd-driven data collection is well-aligned with current technological trends. We also provide some directions about the kind of analytics that can be done on the crowd-collected data in case of different application scenarios. Furthermore, we discuss some of the open research issues in this area.

1 Introduction

Big data analytics for effective decision-making entails the collection of significant amounts of data. Given the increasing prevalence and popularity of mobile devices coupled with the fact that mobile devices often come equipped with various kinds of sensors, crowd-driven data collection is well-aligned with current technological trends. Crowd-driven data collection can facilitate the collection of huge amounts of data at relatively low cost. Furthermore, the involvement of crowdworkers in the data collection process brings in the angle of *human judgment*, thereby creating exciting opportunities for collecting newer types of data and also potentially enabling the filtering out of noisy data during the data collection phase itself.

Crowd-driven data collection can enable data analytics for a wide gamut of applications such as transportation, healthcare, town planning, soil monitoring, agriculture, geographical information systems (GIS) and so on. For realizing crowd-driven data collection in practice, it is of paramount importance to provide incentives to crowdworkers. Incentives can also be tied to reduction of noise in the data, data reliability, data correctness, privacy and so on. The key to encouraging

V. Bhatnagar and S. Srinivasa (Eds.): BDA 2013, LNCS 8302, pp. 86–96, 2013.

crowdworkers to collect data is to provide the 'right' level of incentives [18], and this is a challenging research area in its own right. In this paper, we discuss the role of *incentive-based crowd-driven data collection* towards facilitating big data analytics.

Notably, existing sensor-based data collection mechanisms are typically expensive to deploy due to the high initial fixed costs of installing a large-scale sensor-based system, and sensors also require maintenance, thereby further adding to the costs. Moreover, sensors cannot be cost-effectively installed at all possible locations. Thus, crowd-driven data collection can be a practically viable alternative to sensor-based data collection. However, this does not imply that crowd-driven data collection approaches should replace sensor-based data collection approaches because both approaches have their own advantages. We see the role of crowd-driven data collection as *complementary* to sensor-based data collection.

Various kinds of analytics can be done on the crowd-collected data. For example, in transportation application scenarios, one could use the data for creating prediction models for traffic congestion and hotspots at different points of time during the different days of the week. Since the crowd-collected data would also contain contextual (spatio-temporal) information about events on the road (e.g., traffic accidents, truck breakdowns, temporary road blockades due to rallies/processions and so on), analytics could also be done on the data to facilitate users in planning their trips from location A to location B during a given time-frame.

In a similar vein, in business marketing and sales application scenarios, analytics can be performed on the crowd-collected data concerning the mobility patterns of users and the spatial density (in terms of the number of people per unit area) of various locations in a given city. Such analytics can also be performed to incorporate other factors such as the proximity of a retail store location to bus routes. This can significantly facilitate in planning the location at which a retail store could potentially be opened such that it would generate higher amounts of foot traffic.

Observe that by taking dynamic contextual crowdsourced data about events into consideration, a retail store could also predict its sales revenue for a specific time duration in a fine-grained manner. This would help the retail store in better planning its sales strategy. For example, consider a music store. In the event that a celebrity singer is scheduled to give a performance at the city in which the store is located, this information could be made available to the system from reports submitted by crowdworkers. The system could report this information to the music store, and additionally provide analytics results concerning the extent to which the store's sales could potentially increase due to this event. One might argue that the music store could have found this information on its own by alternative means e.g., by conducting a web search or from news sources. However, this would require the store to have dedicated resources to track this kind of information. By using crowdsourced data, costs can be significantly reduced because the need for deploying dedicated resources for event data collection and tracking would decrease.

Notably, sensor-based technologies do not always support the collection of contextual data (e.g., traffic accidents). Hence, when using sensor-based technologies for data collection, such contextual data has to be obtained by the system in other ways. Furthermore, the contextual data needs to be integrated into the data management system for effective decision-making (e.g., trip planning decisions). In practice, this almost always results in data integration problems due to the contextual data coming from disparate data sources. On the other hand, when using crowd-collected data, contextual data can be almost seamlessly integrated into the system primarily because the system typically receives inputs from the crowd in certain specific formats.

However, many open questions still remain in this research area. The focus of this work is to discuss open research issues concerning the use of incentive-oriented crowdsourcing for data collection, as well as to present our perspectives on these issues. Hence, we will just briefly refer to some of the existing crowdsourcing-based systems and incentives schemes for data management without delving into specific details and as such, this paper is not intended to be a survey.

The remainder of this paper is organized as follows. Section 2 briefly discusses some background information concerning crowd-driven data collection systems as well as incentive approaches, while Section 3 presents our contributions in this area. Section 4 discusses open research issues and our perspectives on these issues. Finally, Section 5 summarizes the paper.

2   Background Information

This section briefly describes some background information concerning the area of incentive-based crowd-driven data collection.

Crowdsourcing-Based Systems for Data Collection

The Amazon Mechanical Turk system (AMT) [2] is a crowdsourcing platform which serves as as an Internet marketplace for performing match-making between tasks that are posted by requestors, and individuals who are qualified to perform those tasks. AMT incorporates crowdworkers' reputation management and provides incentives to crowdworkers for performing the tasks.

The Ushahidi Platform [4] serves as a crowdsourcing tool, which enables data collection in different ways (e.g., via text, email, web and twitter) for spatio-temporal information collection, visualization and interactive mapping. Crowdworkers can submit reports either online or via mobile devices. The capabilities of the Ushahidi platform are complemented by the SwiftRiver Initiative, which provides real-time data collection as well as analytics and intelligence products.

The online FixMyStreet platform [3] enables users to specify the spatial location of any given street on a map, and report problems associated with the street. Examples of problems could be broken pavements, pot-holes and so on. Crowdsourcing systems have also been effective in dealing with a wide gamut of problems such as providing legal assistance to people, helping the governments in exposing illegal spending as well as facilitating city governments in detecting land violations [1].

Economic Models and Incentive Schemes for Data Management

Economic models have been proposed for resource allocation. The proposal in [15] performs resource allocation in distributed systems by incorporating economy-based optimal file allocation. The works in [17,30] discuss economic models for resource allocation in wireless ad hoc networks.

Incentive schemes have been proposed for encouraging contributions from users as well as for combating free-riding. Incentives for static P2P networks use utility functions to capture peer contributions [13] and EigenTrust scores to capture participation criteria [14]. Furthermore, the popular eMule/eDonkey P2P file-sharing network [16] also uses incentive mechanisms to reward contributors.

Incentive schemes have also been proposed to encourage peers to forward messages in mobile ad hoc networks (MANETs) [7,9,26] . These schemes stimulate cooperation among peers by means of virtual currency and counter-based mechanisms at each peer [6]. The work in [29] incentivizes mobile peers towards participation in the dissemination of reports about spatio-temporal resources in Mobile-P2P networks. The proposal in [28] focusses on opportunistic resource information dissemination in transportation application scenarios.

Payment schemes associated with incentives have been proposed for mobile environments. Examples include coupon-based systems such as adPASS [27] and Coupons [12]. The eNcentive framework [25] enables agents to disseminate digital advertisements with embedded coupons. Public-key cryptography and digital-watermarking technology are used to prevent manipulation of the e-coupon [7]. The works in [10,11,31] discuss how to ensure secure payments using a virtual currency.

3   Our Contributions

This section briefly summarizes our contributions in the area of economic incentive models for data collection and management in mobile environments.

It is a well-established fact that the majority of humans respond to incentives. As a single instance, a study [19], which was conducted on users' motivation and decision to share resources in P2P networks, revealed that 50% of the questioned users would share more, if some materialistic incentives (e.g., money) are dispensed by the application. In a similar vein, a large percentage of the peers in P2P environments are typically free-riders [5] i.e., they do not provide any data to the network because there are no incentives for collecting and sharing the data. The same argument of incentives is also applicable to the case for mobile crowdworkers performing data collection. Notably, mobile crowdworkers also need to expend the limited resources (e.g., energy, bandwidth) of their mobile devices, thereby further motivating the need for incentives. This has motivated our contributions in the area of incentives.

The work in [24] discusses economic incentive-based brokerage schemes for facilitating efficient data collection as well as for improving data availability in mobile environments. It also incorporates broker rating strategies for providing additional incentives to brokers towards providing better service in terms of data collection. Here, the brokers help to consolidate the data collected by other crowdworkers. Brokers can be pre-selected based on the application (e.g., for transportation application scenarios, the traffic police could act as information brokers). Alternatively, brokers can be selected from among the crowdworkers by incentivizing them with broker commissions.

The E-ARL system [21] incorporates an economic incentive scheme for adaptive revenue-load-based dynamic replication of data in dedicated mobile networks. It considers a mobile cooperative environment, where the mobile crowdworkers are working towards the same goal (e.g., performing data collection for a specific application), and the network performance is facilitated by the economic scheme. Each type of data item is associated with a price in terms of a virtual currency, and the price is determined by considering various factors such as urgency of use, data availability and popularity and so on. The proposed incentive mechanism can be used effectively for facilitating large-scale data collection by providing different amounts of incentives for the collection of different types of data by the crowdworkers. It also conserves the energy of low-energy mobile crowdworkers to facilitate network connectivity so that the data collection remains sustainable.

The EcoTop system [23] incorporates an economic model, which can facilitate the reduction of noise during data collection. In particular, it issues economic rewards to the mobile crowdworkers, which send relevant data, and penalizes them for sending irrelevant data, thereby incentivizing the optimization of the collected data as well as the communication traffic. Notably, such noise reduction in the collected data has the potential of significantly reducing computational overheads associated with data cleaning, and it also helps in facilitating meaningful analytics on the relatively noise-free data.

The LEASE system [22] proposes a novel mobile-P2P lease-based economic incentive model, in which data requestors need to pay the price (in virtual currency) of their requested data to data-providers, which are mobile crowdworkers. In LEASE, mobile crowdworkers, which collect data, can lease the data to those, who are looking for the data, in lieu of a lease payment. In effect, such a lease model entices even those crowdworkers, who have no data to provide, to host data from other crowdworkers. The system uses a Vickrey auction-based bidding mechanism as part of its incentive scheme, thereby providing incentives to crowdworkers for improving data availability by performing data leasing.

The proposed approach in [20] can be used for data collection in mobile environments, where data collection must adhere to constraints such as data quality and trustworthiness of the data source. It also provides incentives for peer collaboration among the crowdworkers in order to improve data availability. In particular, it provides incentives for crowdworkers to form collaborative peer groups, which can facilitate data collection by mutually allocating and deallocating data among group members using a royalty-based revenue-sharing mechanism. As enablers, it uses Vickrey auctions and it proposes the CR*-tree, which is a dynamic multi-dimensional R-tree-based index for dealing with the constraints of data quality, trust and price.

4   Open Research Issues and Perspectives

This section discusses open questions concerning incentive-based crowd-driven data collection for big data analytics purposes. We also point out our perspectives on these open questions.

Noisy Data

Significant amounts of data are often collected by data management systems without any clear goals concerning the future usage of the data i.e., how the data is intended to be used for decision-making purposes. Consequently, considerable amounts of noise get unnecessarily added to the data in the sense that a large percentage of the collected data does not contribute towards effective decision-making. This poses significant challenges to big data analytics because the noisy data has to be filtered out first before any meaningful analytics can be performed on the data.

To address the issue of noisy data, an effective approach could be to clearly define goals concerning how the data is intended to be used in the future, thereby facilitating the filtering out of the noisy data during the data collection phase itself. Depending upon the granularity of these goals (fine-grained vs coarse-grained), there would be certain amount of remaining noisy data, and such noisy data can be addressed by using existing approaches. Furthermore, research efforts on *learning* the granularity of these goals w.r.t. various application scenarios are necessary to reduce the amounts of noisy data as much as possible. Additionally, the knowledge of domain experts should be leveraged to determine the granularity of such goals.

Noisy data can also arise from the lack of expertise of the crowdworkers in collecting the required data. As a single instance, image data typically needs to be collected in various applications involving transportation. In such scenarios, there could be various factors (e.g., insufficient light conditions, low-resolution camera, the angle at which the photo is taken and so on) that contribute to bad quality photos/images. Here, we need to ensure that only the good-quality images are collected as data by the system, while the bad-quality images are filtered out by the system.

One possible approach to achieve this could be to link incentives to the quality of data that is being collected by crowdworkers. In other words, only the crowdworkers, who provide good-quality photos as data, would receive incentives, while the others would not receive any incentives. This incentivizes crowdworkers towards collecting data, whose quality is above a system-mandated threshold, thereby resulting in the reduction of noise in the data. The effectiveness of this approach could be further improved by training and educating crowdworkers about the best practices in good-quality data collection for various application scenarios. For example, in an image data collection scenario for a given application in transportation, one could train crowdworkers about taking photos in settings involving insufficient light conditions as well as educating them about the correct angles at which the photos would have more clarity.

Data Reliability

Although crowdsourcing enables collection of huge amounts of data at low cost, it does not provide any guarantees of data reliability. It is a well-known fact that performing analytics on unreliable data could generate potentially erroneous insights, thereby leading to bad decision-making. To ensure higher data reliability, one approach could be to incorporate redundancies in the data collection i.e., collect more data than necessary. For example, one could collect data from a large number of mobile users about a traffic congestion scenario to conclusively determine that there is indeed a traffic congestion at a given location during a specific time duration. However, adopting this approach would introduce additional computational overheads in data de-duplication and data cleaning because the system just needs to store the data that there was a traffic congestion at location X during a specific time duration as opposed to storing all the data that led to this conclusion.

Another approach to improve data reliability could be to associate trust ratings with each crowdworker such that crowdworkers with higher trust ratings get better incentives. For example, if a few crowdworkers with high trust ratings report to the system that there has been an accident at a given location, the system records this incident. On the other hand, if the crowdworkers have low trust ratings, the number of crowdworkers reporting the same incident would have to increase correspondingly for the data about the incident to be recorded into the system.

Incentives for Crowd-Driven Data Collection

What kind of incentives entice crowdworkers to collect data and send it to a data analytics system? There could be either intrinsic motivation or extrinsic motivation or both. Despite the significant body of literature on incentive management, getting incentive-based systems to work in practice poses several challenges from the practical perspective. Some key research challenges in this space are as follows:

- What is the 'right' kind of incentives for crowdworkers to collect data? Are the crowdworkers motivated by altruism (intrinsic motivation) or by economics (extrinsic motivation)?

– Does the 'right' kind of incentives change based on context? For example, crowdworkers may be intrinsically motivated to report a serious traffic accident involving several cars because this constitutes an emergency. On the other hand, they may require extrinsic motivation to report incidents that are far less serious from their point of view.
– What is the 'right' level of incentives? This is a fundamentally difficult question since it depends upon several factors such as the demographics, purchasing power etc., of the crowdworkers.
– How easy are these incentives to redeem? From a practical standpoint, incentives have to be such that they are easy for the crowdworker to redeem. Given the high cost of micro-economic transactions, a significant amount of research has been done on examining virtual currency based incentives such as reward points. While such virtual currency based incentives sometimes become necessary to minimize security-related overheads of handling real currency, meticulous care should be taken to ensure that such incentives are easy to redeem, thereby providing value to crowdworkers.

Scalability

Given the huge amounts of data that can potentially be collected by the crowdworkers over a period of time, managing the scale of the data becomes a major issue. Such huge amounts of data can easily overwhelm data storage systems in addition to incurring significant storage costs. Furthermore, given that typically only a small percentage of the data is likely to be useful for analytics purposes, it becomes a necessity to have a filtering mechanism to ensure that the useful data does not get buried under huge amounts of irrelevant data.

In this regard, an interesting issue arises: *Should crowdworkers be penalized in order to deter them from sending irrelevant data to the system?* To adequately address this issue, one has to consider the trade-off between data relevance and crowdworker participation. If the system penalizes crowdworkers for sending irrelevant data, data reliability would be likely to increase, but crowdworker participation would decrease. Consequently, there would also be a decrease in the amount of the crowd-collected data. On the other hand, if no penalties are assigned to the crowdworkers for sending completely irrelevant data, data reliability would be likely to decrease, although the amount of data collected would be high due to increased participation by crowdworkers.

Our perspective on this issue is that the system should try to be inclusive as far as possible. Hence, crowdworkers should never be penalized. The rationale for this recommendation is that any crowdsourcing-based system thrives on participation by the crowdworkers. Thus, alienating the crowdworkers by penalizing them would defeat the very purpose of crowdsourcing.

Expectations Gap between IT Experts and Domain Experts

IT experts often lack domain knowledge. When an IT expert designs a system for collecting crowdsourced data and doing analytics on the data, he may not be able to appreciate the type of analytics results that would actually be useful to a domain expert. Notably, in many of the cases, domain experts are typically the end-users of analytics results.

As an example, let us consider the transportation domain. An IT expert can do significant amount of analytics on transportation data, which has been collected by means of crowdsourcing. As a result, he may find a large number of patterns such as "In location X, traffic congestion is highest during 9 am to 10 am during weekdays", "In location Y, traffic congestion is highest during 2 pm to 5 pm during weekends", and so on. However, many of these analytics results would be already known and probably seem obvious to a domain expert in transportation (and even to users who regularly move around these areas).

It is of paramount importance to understand that one of the key value-additions gained by investing resources into data analytics is to generate insights that a domain expert would not be able to envision. Generating such *interesting* and *unexpected* analytics results is definitely challenging. We believe that close engagement between IT experts and domain experts during the design of the system would be likely to considerably increase the chances of finding interesting analytics results and generating meaningful insights that contribute to effective decision-making. As a single instance, inputs from the domain experts would facilitate the system in distinguishing between the kind of data that the system should store (for analytics purposes) and the kind of data that the system should filter out as irrelevant.

Privacy-Preserving Data Collection

When using crowdsourcing as a means of data collection, privacy is a major issue. For example, when taking photos or videos of traffic congestion scenarios, all users in those photos/videos should be anonymized so that the data collection is privacy-aware. Furthermore, fear of abuse of users' location information by centralized location-dependent service providers may hinder the growth of pervasive applications that require participative sensing by mobile devices [8]. However, this is a challenging problem in practice simply because most crowd-workers may not be sensitive as far as privacy issues are concerned. One possible approach could be to educate crowdworkers to make them more aware about privacy issues, but this is easier said than done. In essence, privacy-awareness is necessary to ensure that crowd-driven data collection remains sustainable over the long-term.

5 Conclusion

This paper has examined the role of incentive-based crowd-driven data collection, and provided some insights about the kind of analytics that can be done on the crowd-collected data. Open research issues and perspectives have been discussed with the objective of soliciting contributions in this area from academia as well as from industry, the final aim being to ensure the collection of high-quality data by means of crowdsourcing approaches.

References

1. http://innovation.internews.org/blogs/opportunities-crowdsourcing-russia-and-ukraine
2. Amazon Mechanical Turk, http://www.mturk.com/mturk/welcome
3. FixMyStreet Platform, http://www.fixmystreet.com/
4. Ushahidi Platform, http://www.ushahidi.com/products/ushahidi-platform
5. Adar, E., Huberman, B.A.: Free riding on Gnutella. Proc. First Monday 5(10) (2000)
6. Buttyan, L., Hubaux, J.P.: Stimulating cooperation in self-organizing mobile ad hoc networks. ACM/Kluwer Mobile Networks and Applications 8(5) (2003)
7. Chen, K., Nahrstedt, K.: iPass: an incentive compatible auction scheme to enable packet forwarding service in MANET. In: Proc. ICDCS (2004)
8. Cornelius, C., Kapadia, A., Kotz, D., Peebles, D., Shin, M., Triandopoulos, N.: Anonysense: privacy-aware people-centric sensing. In: Proc. MobiSys (2008)
9. Crowcroft, J., Gibbens, R., Kelly, F., Ostring, S.: Modelling incentives for collaboration in mobile ad hoc networks. In: Proc. WiOpt (2003)
10. Daras, P., Palaka, D., Giagourta, V., Bechtsis, D.: A novel peer-to-peer payment protocol. In: Proc. IEEE EUROCON (2003)
11. Elrufaie, E., Turner, D.: Bidding in P2P content distribution networks using the lightweight currency paradigm. In: Proc. ITCC (2004)
12. Garyfalos, A., Almeroth, K.C.: Coupon based incentive systems and the implications of equilibrium theory. In: Proc. IEEE Conference on E-Commerce Technology (2004)
13. Ham, M., Agha, G.: ARA: A Robust Audit to prevent free-riding in P2P networks. In: Proc. P2P (2005)
14. Kamvar, S.D., Schlosser, M.T., Garcia-Molina, H.: Incentives for combatting free-riding on P2P networks. In: Kosch, H., Böszörményi, L., Hellwagner, H. (eds.) Euro-Par 2003. LNCS, vol. 2790, pp. 1273–1279. Springer, Heidelberg (2003)
15. Kurose, J.F., Simha, R.: A microeconomic approach to optimal resource allocation in distributed computer systems. IEEE Trans. Computers 38(5) (1989)
16. Li, Y., Gruenbacher, D., Scoglio, C.M.: Reward only is not enough: Evaluating and improving the fairness policy of the P2P file sharing network eMule/eDonkey. Journal of Peer-to-Peer Networking and Applications 5(1) (2012)
17. Liu, J., Issarny, V.: Service allocation in selfish mobile ad hoc networks using vickrey auction. In: Lindner, W., Fischer, F., Türker, C., Tzitzikas, Y., Vakali, A.I. (eds.) EDBT 2004. LNCS, vol. 3268, pp. 385–394. Springer, Heidelberg (2004)
18. Madria, S.K., Mondal, A.: Economic-based Incentive Schemes for Dynamic Data Management in Mobile P2P Computing. In: Proc. MDM (2008)
19. Mannak, R., de Ridder, H., Keyson, D.V.: The human side of sharing in peer-to-peer networks. In: Proc. ACM EUSAI (2004)
20. Mondal, A., Madria, S.K., Kitsuregawa, M.: An economic incentive model for encouraging peer collaboration in mobile-P2P networks with support for constraint queries. Journal of Peer-to-Peer Networking and Applications (2009)
21. Mondal, A., Madria, S.K., Kitsuregawa, M.: E-ARL: An Economic incentive scheme for Adaptive Revenue-Load-based dynamic replication of data in Mobile-P2P networks. Journal of Distributed and Parallel Databases 28 (2010)
22. Mondal, A., Madria, S.K., Kitsuregawa, M.: Improving data availability via an economic lease model in M-P2P networks. Journal of Computer Science and Engineering 5 (2010)

23. Padhariya, N., Mondal, A., Goyal, V., Shankar, R., Madria, S.K.: EcoTop: an economic model for dynamic processing of top-k queries in mobile-P2P networks. In: Yu, J.X., Kim, M.H., Unland, R. (eds.) DASFAA 2011, Part II. LNCS, vol. 6588, pp. 251–265. Springer, Heidelberg (2011)
24. Padhariya, N., Mondal, A., Madria, S.K., Kitsuregawa, M.: Economic incentive-based brokerage schemes for improving data availability in mobile-P2P networks. Journal of Computer Communications 36(8) (2013)
25. Ratsimor, O., Finin, T., Joshi, A., Yesha, Y.: eNcentive: A framework for intelligent marketing in Mobile Peer-to-Peer environments. In: Proc. ICEC (2003)
26. Srinivasan, V., Nuggehalli, P., Chiasserini, C.F., Rao, R.R.: Cooperation in wireless ad hoc networks. In: Proc. INFOCOM (2003)
27. Straub, T., Heinemann, A.: An anonymous bonus point system for mobile commerce based on word-of-mouth recommendation. In: Proc. ACM SAC (2004)
28. Wolfson, O., Xu, B., Sistla, A.P.: An economic model for resource exchange in Mobile Peer-to-Peer networks. In: Proc. SSDBM (2004)
29. Xu, B., Wolfson, O., Rishe, N.: Benefit and pricing of spatio-temporal information in Mobile Peer-to-Peer networks. In: Proc. HICSS-39 (2006)
30. Xue, Y., Li, B., Nahrstedt, K.: Optimal resource allocation in wireless ad hoc networks: A price-based approach. IEEE Transactions on Mobile Computing (2005)
31. Zhong, S., Chen, J., Yang, Y.R.: Sprite: A simple, cheat-proof, credit-based system for mobile ad-hoc networks. In: Proc. INFOCOM (2003)

Discovering Quasi-Periodic-Frequent Patterns in Transactional Databases

R. Uday Kiran and Masaru Kitsuregawa

Institute of Industrial Science,
The University of Tokyo, Tokyo, Japan
{uday_rage,kitsure}@tkl.iis.u-tokyo.ac.jp

Abstract. Periodic-frequent patterns are an important class of user-interest-based frequent patterns that exist in a transactional database. A frequent pattern can be said *periodic-frequent* if it appears periodically throughout the database. We have observed that it is difficult to mine periodic-frequent patterns in very large databases. The reason is that the occurrence behavior of the patterns can vary over a period of time causing periodically occurring patterns to be non-periodic and/or vice-versa. We call this problem as the "intermittence problem." Furthermore, in some of the real-world applications, the users may be interested in only those frequent patterns that might have appeared almost periodically throughout the database. With this motivation, we relax the constraint that a pattern must appear periodically throughout the database, and introduce a new class of user-interest-based frequent patterns, called quasi-periodic-frequent patterns. Informally, a frequent pattern is said to be *quasi-periodic-frequent* if most of its occurrences are periodic in a database. We propose a model and a pattern-growth algorithm to discover these patterns. The proposed patterns do not satisfy the downward closure property. We have introduced three pruning techniques to reduce the computational cost of mining the patterns. Experimental results show that the proposed patterns can provide useful information and the proposed algorithm is efficient.

Keywords: Data mining, knowledge discovery in databases, frequent patterns and periodic behavior.

1 Introduction

Periodic-frequent patterns [16] are an important class of regularities that exist within a transactional database. In many real-world applications, these patterns can provide useful information pertaining to those patterns that are not only occurring frequently, but also appearing at regular intervals specified by the user in a database. For example, in a retail market, among all frequently sold products, the user may be interested only in the regularly sold products compared to the rest. The application of periodic-frequent pattern mining on a market basket data may provide such useful information to the users. The basic model of periodic-frequent patterns is as follows [16].

Let $I = \{i_1, i_2, \cdots, i_n\}$ be a set of items. A set $X = \{i_1, \cdots, i_k\} \subseteq I$, where $1 \leq k \leq n$ is called a **pattern** (or an itemset). A pattern containing k number of items is called k-**pattern**. A transaction $t = (tid, Y)$ is a tuple, where tid represents a transaction-id (or a

V. Bhatnagar and S. Srinivasa (Eds.): BDA 2013, LNCS 8302, pp. 97–115, 2013.

timestamp) and Y is a pattern. A transactional database T over I is a set of transactions, $T = \{t_1, \cdots, t_m\}$, $m = |T|$, where $|T|$ is the size of T in total number of transactions. If $X \subseteq Y$, it is said that t contains X (or X occurs in t) and such transaction-id is denoted as t_j^X, $j \in [1, m]$. Let $T^X = \{t_k^X, \cdots, t_l^X\} \subseteq T$, where $k \leq l$ and $k, l \in [1, m]$ be the ordered set of transactions in which the pattern X has occurred in the database. The **support** of X, denoted as $S(X)$, represents the number of transactions containing X in T. That is, $S(X) = |T^X|$. The pattern X is said to be **frequent** if $S(X) \geq minsup$, where $minsup$ is the user-defined minimum support threshold value. Let t_j^X and t_{j+1}^X, where $j \in [k, (l-1)]$ be two consecutive transactions in T^X. The number of transactions (or the time difference) between t_{j+1}^X and t_j^X can be defined as a **period** of X, say p_a^X. That is, $p_a^X = t_{j+1}^X - t_j^X$. Let $P^X = \{p_1^X, p_2^X, \cdots, p_r^X\}$, $r = |T^X| + 1$, be the complete set of periods for pattern X. The **periodicity** of X, denoted as $Per(X)$, represents the maximum period of X in T. That is, $Per(X) = max(p_1^X, p_2^X, \cdots, p_r^X)$. The frequent pattern X can be said **periodic-frequent** if $Per(X) \leq maxprd$, where $maxprd$ refers to the user-specified maximum periodicity threshold value. Please note that both the *support* and *periodicity* of a pattern can be described in percentage of $|T|$.

TID	Items	TID	Items
1	a,b	6	e,f
2	c,d	7	c,d
3	a,b,e,f	8	a,b
4	b,e	9	c,d,e,f
5	a,b,c,d	10	a,b

a) **Transactional database**   b) **Pattern occurrences**

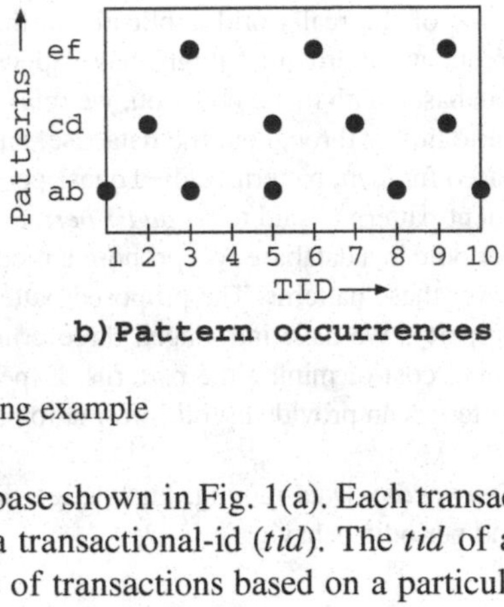

Fig. 1. Running example

Example 1. Consider the transactional database shown in Fig. 1(a). Each transaction in this database is uniquely identifiable with a transactional-id (*tid*). The *tid* of a transaction also represents the ordered sequence of transactions based on a particular time stamp. **In this database, let us consider a sub-database consisting of the first five transactions.** The set of items, $I = \{a, b, c, d, e, f\}$. The set of 'a' and 'b' i.e., $\{a, b\}$ is a pattern. For the purpose of simplicity, we represent this pattern as 'ab'. This pattern occurs in the *tids* of 1, 3 and 5. Therefore, $T^{ab} = \{1, 3, 5\}$. The support of 'ab', i.e., $S(ab) = |T^{ab}| = 3$. If the user-specified $minsup = 2$, then 'ab' is a frequent pattern because $S(ab) \geq minsup$. The periods for this pattern are $p_1^{ab} = 1 (= 1 - t_i)$, $p_2^{ab} = 2 (= 3 - 1)$, $p_3^{ab} = 2 (= 5 - 3)$ and $p_4^{ab} = 0 (= t_l - 5)$, where $t_i = 0$ represents the initial transaction and $t_l = 5$ represents the last transaction in the sub-transactional database. The periodicity of 'ab', $Per(ab) = maximum(1, 2, 2, 0) = 2$. If the user-specified $maxprd = 2$, then the frequent pattern 'ab' is also a periodic-frequent pattern because $Per(ab) \leq maxprd$.

The time duration of a database refers to the difference between the starting and ending timestamps of the transactions within it. The time duration of a database is independent to its size (i.e., number of transactions within it) because arrival rates of the transactions can vary with respect to time. We have observed that it is difficult to mine periodic-frequent patterns if a database is composed over a very long time duration. It is because items' occurrence behavior can vary drastically causing many periodically occurring frequent patterns to be non-periodic and/or vice-versa.

Example 2. In a shopping mall, if we consider the transactions happened only during the winter season, then we may find the interesting periodic-frequent patterns pertaining to the woolen wears. However, if we consider the transactions of the whole year, then it might be difficult for the user to find the periodic-frequent patterns pertaining to the woolen wears. It is because the woollen wears are not often purchased during the summer season.

We call this problem as the "*intermittence problem.*" A method to address this problem involves mining the patterns with a high *maxprd* value. However, such an approach may generate sporadically occurring frequent patterns as the periodic-frequent patterns.

Example 3. Let us consider the complete transactional database shown in Fig. 1(a).The pattern '*ab*' occurs in *tids* of $1, 3, 5, 8$ and 10. Therefore, $T^{ab} = \{1, 3, 5, 8, 10\}$. The support of '*ab*,' i.e., $S(ab) = |T^{ab}| = 5$. The periods for this pattern are $p_1^{ab} = 1 (= 1 - t_i)$, $p_2^{ab} = 2 (= 3 - 1)$, $p_3^{ab} = 2 (= 5 - 3)$, $p_4^{ab} = 3 (= 8 - 5)$, $p_5^{ab} = 2 (= 10 - 8)$ and $p_6^a = 0 (= t_l - 10)$. The set of periods for the pattern '*ab*', $P^{ab} = \{1, 2, 2, 3, 2, 0\}$. The *periodicity* of '*ab*', i.e., $per(ab) = max(1, 2, 2, 3, 2, 0) = 3$. If the user-specified *maxprd* $= 2$, then '*ab*' is a non-periodic frequent pattern because $Per(ab) > maxprd$. We can discover the pattern '*ab*' as a periodic-frequent pattern by setting a high *maxprd* value, say *maxprd* $= 3$. However, this high *maxprd* value can result in generating the sporadically occurring frequent pattern, say '*ef*', as a periodic-frequent pattern because $Per(ef) = 3$ (see Fig. 1(b)).

With this motivation, we have investigated the interestingness of the frequent patterns with respect to their proportion of periodic occurrences in the database. During this investigation, we have observed that the frequent patterns that were almost occurring periodically in a database can also provide useful knowledge to the users. Based on this observation, we relax the constraint that a frequent pattern must appear periodically throughout the database and introduce a new class of user-interest-based frequent patterns, called quasi-periodic-frequent patterns. Informally, a frequent pattern is said to be *quasi-periodic-frequent* if most of its occurrences are periodic in a transactional database. Please note that the proposed patterns are tolerant to the intermittence problem. The contributions of this paper are as follows:

- An alternative interestingness measure, called periodic-ratio, has been proposed to assess the periodic interestingness of a pattern.
- The quasi-periodic-frequent patterns do not satisfy the *downward closure property*. That is, all non-empty subsets of a quasi-periodic-frequent pattern may not be quasi-periodic-frequent. This increases the search space, which in turn increases the computational cost of mining the patterns. We introduce three pruning techniques to reduce the computational cost of mining the patterns.
- A pattern-growth algorithm, called Quasi-Periodic-Frequent pattern-growth (QPF-growth), has been proposed to discover the patterns.

– Experimental results show that the proposed patterns can provide useful information, and the QPF-growth is efficient.

The rest of the paper is organized as follows. Section 2 describes the efforts made in the literature to discover periodic-frequent patterns. In Section 3, we present the conceptual model of quasi-periodic-frequent patterns. In Section 4, we discuss the basic idea to reduce the computational cost of mining the patterns and describe the QPF-growth algorithm. Experimental results are presented in Section 5. Section 6 concludes the paper with the future research directions.

2   Related Work

The periodic behavior of patterns has been widely studied in various domains as temporal patterns [3], cyclic association rules [13] and periodic behavior of moving objects [11]. Since real-life patterns are generally imperfect, Han et al. [6] have investigated the concept of partial periodic behavior of patterns in time series databases with the assumption that period lengths are known in advance. The **partial periodic patterns** specify the behavior of time series at some but not at all points in time. Sheng and Joseph [12] have investigated the same with unknown periods. Walid et al. [4] have extended the Han's work in [6] to incremental mining of partial periodic patterns in time series databases. All of these approaches do not consider the effect of noise on the partial periodic behavior of patterns. Therefore, Yan et al. [17] have extended [6] by introducing information theory concepts to address the effect of noise on periodic behavior of patterns in time series. Sheng et al. [15] have introduced the concept of density to discover partial periodic patterns that may have occurred within small segments of time series data. The problem of mining frequently occurring periodic patterns with a gap has been investigated in sequence databases [10,18,19]. Although these works are closely related to our work, they cannot be directly applied for finding quasi-periodic-frequent patterns from a transactional database because of two reasons. First, they consider either time series or sequential data; second, they do not consider the support threshold which needs to be satisfied by all frequent patterns.

Recently, the periodic occurrences of frequent patterns in a transactional database has been studied in the literature to discover periodic-frequent patterns [9,14,16]. These patterns suffer from the "intermittence problem." Furthermore, these approaches cannot be directly extended to discover the quasi-periodic-frequent patterns. The reason is that the proposed patterns do not satisfy the downward closure property.

The concept of finding frequent patterns in data streams uses the temporal occurrence of a pattern in a database [5]. However, it has to be noted that only the occurrence frequency was used to discover frequent patterns.

3   Proposed Model

In this section, we describe the model of quasi-periodic-frequent patterns using the periodic-frequent pattern mining model discussed in Section 1. We also introduce the basic notations and definitions in this regard.

Definition 1. Maximum period (*maxperiod*): *It is a user-specified constraint that describes the interestingness of a period. A period of X, $p_a^X \in P^X$, is interesting if $p_a^X \leq maxperiod$.*

Example 4. Continuing with the Example 3, if the user-specified *maxperiod* = 2, then $p_1^{ab}, p_2^{ab}, p_3^{ab}, p_5^{ab}$ and p_6^{ab} are the interesting periods because their values are less than or equal to the user-specified *maxperiod* constraint.

Definition 2. *The periodic-ratio of a pattern X:* *Let $IP^X = \{p_i^X, \cdots, p_j^X\} \subseteq P^X, 0 \leq i \leq j \leq r$, be the set of interesting periods of the pattern X. That is, $\forall p_j^X \in IP^X, p_j^X \leq$ maxperiod. The* periodic-ratio *of pattern X, denoted as $pr(X) = \frac{|IP^X|}{|P^X|}$. This measure captures the proportion of periods in which pattern X has appeared periodically in a database.*

For a pattern X, $pr(X) \in [0, 1]$. If $pr(X) = 0$, it means X has not appeared periodically anywhere in the transactional database. If $pr(X) = 1$, it means X has appeared periodically throughout the transactional database. In other words, X is a periodic-frequent pattern. The proportion of non-periodic occurrences of a pattern X in the transactional database is given by $1 - pr(X)$. Please note that the *periodic-ratio* of a pattern X can be measured in percentage of P^X.

Example 5. Continuing with the Example 4, $IP^{ab} = \{p_1^{ab}, p_2^{ab}, p_3^{ab}, p_5^{ab}, p_6^{ab}\}$. Therefore, the periodic-ratio of '*ab*,' i.e., $pr(ab) = \frac{|IP^{ab}|}{|P^{ab}|} = \frac{5}{6} = 0.833 (= 83.3\%)$. The periodic-ratio of '*ab*' says that 0.833 proportion of its occurrences are periodic and 0.167 (=1-0.833) proportion of its occurrences are non-periodic within the database.

Definition 3. Minimum periodic-ratio (*minpr*): *It is a user-specified constraint that describes the interestingness of a pattern with respect to its periodic occurrences in a database. The pattern X is periodically interesting if $pr(X) \geq minpr$.*

Example 6. Continuing with the Example 5, if the user-specified *minpr* = 0.8, then the pattern '*ab*' is interesting with respect to its periodic occurrences, because, $pr(ab) \geq minpr$.

Definition 4. *Quasi-periodic-frequent pattern:* *The pattern X is said to be quasi-periodic-frequent if $S(X) \geq minsup$ and $pr(X) \geq minpr$. For a quasi-periodic-frequent pattern X, if $S(X) = a$ and $pr(X) = b$, then it is described as shown in Equation 1.*

$$X \ [support = a, periodic\text{-}ratio - b] \tag{1}$$

Example 7. Continuing with the Example 6, if the user-specified *minsup* = 3, *maxperiod* = 2 and *minpr* = 0.8, then '*ab*' is a quasi-periodic-frequent pattern because $S(ab) \geq minsup$ and $pr(ab) \geq minpr$. This pattern is described as follows:

$$ab \ [support = 5, \ periodic\text{-}ratio = 0.833]$$

The quasi-periodic-frequent patterns mined using the proposed model do not satisfy the *downward closure property*. That is, all non-empty subsets of a quasi-periodic-frequent pattern may not be quasi-periodic-frequent. The correctness of our argument is based on the Properties 1 and 2, and shown in Lemma 1.

Property 1. In the proposed model, the total number of periods for X i.e., $|P^X| = S(X) + 1$.

Property 2. Let X and Y be the two patterns in a transactional database. If $X \subset Y$, then $|P^X| \geq |P^Y|$ and $|IP^X| \geq |IP^Y|$ because $T^X \supseteq T^Y$.

Lemma 1. *In a transactional database T, the quasi-periodic-frequent patterns mined using minsup, maxperiod and minpr constraints do not satisfy the* downward closure *property.*

Proof. Let $Y = \{i_a, \cdots, i_b\}$, where $1 \leq a \leq b \leq n$ be a quasi-periodic-frequent pattern with $S(Y) = minsup$ and $periodic\text{-}ratio(Y) = minpr$. Let Y be an another pattern such that $X \subset Y$. From Property 2, we derive $|P^X| \geq |P^Y|$ and $|IP^X| \geq |IP^Y|$. Considering the scenario where $|P^X| > |P^Y|$ and $|IP^X| = |IP^Y|$, we derive $periodic\text{-}ratio(X) < periodic\text{-}ratio(Y)(= minpr)$. Hence, X is not a quasi-periodic-frequent pattern. Hence proved.

Problem Definition. Given a transactional database T, minimum support ($minsup$), maximum period ($maxperiod$) and minimum periodic-ratio ($minpr$) constraints, the objective is to discover the complete set of quasi-periodic-frequent patterns in T that have support and periodic-ratio no less than the user-defined $minsup$ and $minpr$, respectively.

4   QPF-Growth: Idea, Design, Construction and Mining

4.1   Basic Idea

We have observed two issues while mining quasi-periodic-frequent patterns in the transactional databases. Now, we discuss each of these issues and our approaches to address the same.

Issue 1: The quasi-periodic-frequent patterns mined using the proposed model do not satisfy the *downward closure property*. This increases the search space, which in turn increases the computational cost of mining mining these patterns. To reduce the computational cost of mining these patterns space, we employ the following two pruning techniques.

1. The *minsup* constraint satisfies the *downward closure property* [2]. Therefore, if a pattern X does not satisfy the *minsup*, then X and its supersets can be pruned because they cannot generate any quasi-periodic-frequent pattern.
2. Another interesting idea is as follows. Every quasi-periodic-frequent pattern will have support greater than or equal to *minsup*. Hence, every quasi-periodic-frequent pattern will have at least $(minsup + 1)^1$ number of periods (Property 1). That is, for a pattern X, $|P^X| \geq (minsup + 1)$. Therefore, if $\frac{|IP^X|}{(minsup+1)} < minpr$, then X cannot be quasi-periodic-frequent pattern. If $Y \supset X$, then Y cannot be a quasi-periodic-frequent pattern because $|IP^Y| \leq |IP^X|$ and $|P^X| \geq |P^Y| \geq (minsup + 1)$. Therefore, if $\frac{|IP^X|}{(minsup+1)} < minpr$, then X and all of its supersets can be pruned because none of them can generate a quasi-periodic-frequent pattern.

---

[1] In this paper, *minsup* is discussed in terms of minimum support count.

Example 8. Let the user-specified *minsup, maxperiod* and *minpr* values for the transactional database shown in Fig. 1(a) be 3, 2 and 0.8, respectively. Every quasi-periodic-frequent pattern will have support greater than or equal 3. Thus, every quasi-periodic-frequent pattern will have at least $4(=3+1)$ number of periods (Property 1). For item 'e', $T^e = \{3,4,6,9\}$, $P^e = \{3,1,2,3,1\}$ and $IP^e = \{1,2,1\}$. Therefore, $\frac{|IP^e|}{minsup+1} = \frac{3}{4} = 0.75 < minpr$. The item '$e$' cannot be quasi-periodic-frequent because *periodic-ratio*$(e) < minpr$ as $|P^e| \geq (minsup+1)$. For any superset of 'e', say 'ef', to be quasi-periodic-frequent, it should also have at least 4 number of periods. Using Property 2, we derive $T^{ef} \subseteq T^e$, $|P^{ef}| \leq |P^e|$ and $|IP^{ef}| \leq |IP^e|$. Since $|IP^{ef}| \leq |IP^e|$, it turns out that $\frac{|IP^{ef}|}{minsup+1} \leq \frac{|IP^e|}{minsup+1} < minpr$. Therefore, '$ef$' is also not a quasi-periodic-frequent pattern. In other words, all supersets of 'e' are also not quasi-periodic-frequent.

Definition 5. *Mai of X: Let IP^X be the set of interesting periodic occurrences for pattern X. The mai of X, denoted as $mai(X) = \frac{|IP^X|}{(minsup+1)}$.*

For a pattern X, $mai(X) \in [0, \infty)$. If $mai(X) \geq 1$, then $S(X) \geq minsup$ because $IP^X \subseteq P^X$. If $mai(X) < 1$, then $S(X) \geq minsup$ or $S(X) < minsup$. So, we have to consider both *mai* and support values of a pattern for generating quasi-periodic-frequent patterns. A **pruning technique** is as follows. If $mai(X) < minpr$, then X and its supersets cannot generate any quasi-periodic-frequent pattern. However, if $mai(X) \geq minpr$ and *periodic-ratio*$(X) < minpr$, then X must be considered for generating higher order patterns even though X is not a quasi-periodic-frequent pattern. The reason is that its supersets can still be quasi-periodic-frequent.

Definition 6. *Potential pattern: For a pattern X, if $mai(X) \geq minpr$ and $S(X) \geq minsup$, then X is said to be a potential pattern.*

A potential pattern containing only one item (1-itemset) is called a **potential item**. A potential pattern need not necessarily be a quasi-periodic-frequent pattern. However, every quasi-periodic-frequent pattern is a potential pattern. Therefore, we employ potential patterns to discover the quasi-periodic-frequent patterns.

Issue 2: The conventional (frequent) pattern-growth algorithms that are based on FP-tree [7] cannot be used to discover these patterns. It is because the structure of FP-tree captures only the occurrence frequency and disregards the periodic behavior of a pattern. To capture both frequency and periodic behavior of the patterns, an alternative pattern-growth algorithm based on a tree structure, called Periodic-Frequent tree (PF-tree), has been proposed in the literature [16]. The nodes in PF-tree do not maintain the support count as in FP-tree. Instead, they maintain a list of *tids* (or a *tid*-list) in which the corresponding item has appeared in a database. These *tid*-lists are later aggregated to derive the final *tid*-list of a pattern (i.e., T^X for pattern X). A complete search on this *tid*-list gives the *support* and *periodic-ratio*, which are later used to determine whether the corresponding pattern is quasi-periodic-frequent or a non-quasi-periodic-frequent pattern. In other words, the pattern-growth technique has to perform a complete search on a pattern's *tid*-list to determine whether it is quasi-periodic-frequent or a non-quasi-periodic-frequent pattern.

In very large databases, the *tid*-list of a pattern can be very long. In such cases, the task of performing a complete search on a pattern's *tid*-list can be a computationally expensive process. To reduce the computational cost, we introduce another pruning technique based on the greedy search. The technique is as follows:

Let $\widehat{P^X} = \{p_1^X, p_2^X, \cdots, p_k^X\}$, $k \le |P^X|$, be the set of periods such that $\widehat{P^X} \subseteq P^X$. Let $\widehat{IP^X} \subseteq \widehat{P^X}$ be the another of set of periods such that $\forall p_k^X \in \widehat{IP^X}$, $p_k^X \le maxperiod$. If $|\widehat{P^X}| > |P^X| \times (1 - minpr)$ and $|\widehat{IP^X}| = 0$, then X is not a quasi-periodic-frequent pattern. The $\widehat{P^X}$ and $\widehat{IP^X}$ denote the set of periods and interesting periods discovered from T^X by applying greedy search technique.

The correctness of this technique is shown in Lemma 2. Example 9 illustrates the proposed pruning technique.

Example 9. In a transactional database containing 20 transactions, let an item '*u*' has appeared in the *tids* of 4, 8, 12, 16 and 18. Therefore, $T^u = \{4, 8, 12, 16, 18\}$. The periods of '*u*' are as follows: $p_1^u = 4 \ (4 - t_i)$, $p_2^u = 4 \ (= 8 - 4)$, $p_3^u = 4 \ (= 12 - 8)$, $p_4^u = 4 \ (= 16 - 12)$, $p_5^u = 2 \ (= 18 - 16)$ and $p_6^u = 2 \ (= t_l - 18)$. The set of all periods for '*u*', i.e., $P^u = \{4, 4, 4, 4, 2, 2\}$. The total number of periods in P^u, i.e., $|P^u| = |T^u| + 1 = 5 + 1 = 6 \ (= |\{4, 4, 4, 4, 2, 2\}|)$. If the $minpr = 0.5$, then '*u*' can be a quasi-periodic-frequent pattern if and only if at least $3 \ (= |P^u| \times minpr)$ out of 6 periods in P^u have value no more than the *maxperiod*. Now, let us consider the following four scenarios:

1. If $\widehat{P^u} = \{p_1^u\}$, then $\widehat{IP^u} = \emptyset$ or $|\widehat{IP^u}| = 0$. It is because $p_1^u > maxprd$. At this point, we cannot say that '*u*' is not a quasi-periodic-frequent pattern. The reason is that any 3 among the remaining 5 periods (i.e., p_2^u, p_3^u, p_4^u, p_5^u and p^6) can have values less than or equal to the *maxperiod* threshold.

2. If $\widehat{P^u} = \{p_1^u, p_2^u\}$, then $\widehat{IP^u} = \emptyset$ or $|\widehat{IP^u}| = 0$. Even now, we cannot say that '*u*' is not a quasi-periodic-frequent pattern. It is because any 3 among 4 periods (i.e., p_3^u, p_4^u, p_5^u and p^6) can have values no more than the *maxperiod* threshold.

3. If $\widehat{P^u} = \{p_1^u, p_2^u, p_3^u\}$, then $\widehat{IP^u} = \emptyset$. Still we cannot say that '*u*' is not a quasi-periodic-frequent pattern. It is because the remaining all 3 periods (i.e., p_4^u, p_5^u and p^6) can have values less than or equal to the *maxperiod* threshold.

4. If $\widehat{P^u} = \{p_1^u, p_2^u, p_3^u, p_4^u\}$, then $\widehat{IP^X} = \emptyset$ or $|\widehat{IP^X}| = 0$. Now, we can say that '*u*' is not a quasi-periodic-frequent pattern. It is because '*u*' can at most occur periodically in only 2 periods (i.e., p_5^u and p^6), which is less than 3. Therefore, if $|\widehat{P^u}| > |P^u| \times (1 - minpr)$ and $|\widehat{IP^u}| = 0$, then '*u*' is a non-quasi-periodic-frequent pattern.

Lemma 2. *Let* $\widehat{P^X} = \{p_1^X, p_2^X, \cdots, p_k^X\}$ *be the set of periods such that* $\widehat{P^X} \subseteq P^X$. *Let* $\widehat{IP^X} \subseteq \widehat{P^X}$ *be the another of set of periods such that* $\forall p_k^X \in \widehat{IP^X}$, $p_k^X \le maxperiod$. *If* $|\widehat{P^X}| > |P^X| \times (1 - minpr)$ *and* $|\widehat{IP^X}| = 0$, *then* X *is not a quasi-periodic-frequent pattern.*

Proof. The total number of *periods* for pattern X, i.e., $|P^X| = (S(X) + 1)$ (see Property 1). If a frequent pattern X is quasi-periodic-frequent, then

$$pr(X) \geq minpr$$

$$= \frac{|IP^X|}{|P^X|} \geq minpr$$

$$= |IP^X| \geq |P^X| \times minpr \tag{2}$$

The number of non-periodic occurrences of X is

$$= |P^X| - |IP^X| \tag{3}$$

Considering $|IP^X| = |P^X| \times minpr$, Equation 3 can be written as

$$= |P^X| \times (1 - minpr) \tag{4}$$

Therefore, if $|\widehat{P^X}| > |P^X| \times (1 - minpr)$ and $|\widehat{IP^X}| = 0$, then X is not a quasi-periodic-frequent pattern. Hence proved.

Using the above ideas, the proposed QPF-growth approach discovers quasi-periodic-frequent patterns by constructing a tree structure, called Quasi-Periodic-Frequent tree (QPF-tree). Now, we discuss the structure of QPF-tree.

4.2 Structure of QPF-Tree

The QPF-tree consists of two components: QPF-list and a prefix-tree. QPF-list is a list with three fields: item (i), support or frequency (s) and number of interesting periods (ip). The node structure of prefix-tree in QPF-tree is same as the prefix-tree in PF-tree [16], which is as follows.

The prefix-tree in QPF-tree explicitly maintains the occurrence information for each transaction in the tree structure by keeping an occurrence transaction-id list, called *tid-list*, only at the last node of every transaction. Two types of nodes are maintained in a QPF-tree: ordinary node and *tail*-node. The ordinary node is similar to the nodes used in FP-tree, whereas the latter is the node that represents the last item of any sorted transaction. The structure of a *tail*-node is $N[t_1, t_2, \cdots, t_n]$, where N is the node's item name and $t_i, i \in [1, n]$, (n be the total number of transactions from the root up to the node) is a transaction-id where item N is the last item. Like the FP-tree [2], each node in a QPF-tree maintains parent, children, and node traversal pointers. However, irrespective of the node type, no node in a QPF-tree maintains support count value in it.

The QPF-growth employs the following three steps to discover quasi-periodic-frequent patterns.

1. Construction of QPF-list to identify potential items
2. Construction of QPF-tree using the potential items
3. Mining quasi-periodic-frequent patterns from QPF-tree

We now discuss each of these steps using the transactional database shown in Fig. 1(a). Let the user-specified *minsup*, *maxperiod* and *minpr* values be 3, 2 and 0.8, respectively.

4.3 Construction of QPF-List

Let id_l be a temporary array that explicitly records the $tids$ of the last occurring transactions of all items in the QPF-list. Let t_{cur} be the tid of current transaction. The QPF-list is, therefore, maintained according to the process given in Algorithm 1.

In Fig. 2 we show how the QPF-list is populated for the transactional database shown in Fig. 1(a). With the scan of the first transaction $\{a,b\}$ (i.e., $t_{cur} = 1$), the items 'a' and 'b' in the list are initialized as shown in Fig. 2(a) (lines 4 to 6 in Algorithm 1). The scan on the next transaction $\{c,d\}$ with $t_{cur} = 2$ initializes the items 'c' and 'd' in QPF-list as shown in Fig. 2(b). The scan on next transaction $\{a,b,e,f\}$ with $t_{cur} = 3$ initializes QPF-list entries for the items 'e' and 'f' with $id_l = 3$, $s = 1$ and $ip = 0$ because $t_{cur} > maxperiod$ (line 6 in Algorithm 1). Also, the $\{s; ip\}$ and id_l values for the items 'a' and 'b' are updated to $\{2; 2\}$ and 3, respectively (lines 8 to 11 in Algorithm 1). The QPF-list resulted after scanning third transaction is shown in Fig. 2(c). The QPF-list after scanning all ten transactions is given in Fig. 2(d). To reflect the correct number of interesting periods for each item in the QPF-list, the whole QPF-list is refreshed as mentioned from lines 16 to 20 in Algorithm 1. The resultant QPF-list is shown in Fig. 2(e). Based on the above discussed ideas, the items 'e' and 'f' are pruned from the QPF-list because their mai values are less than $minpr$ (lines 22 to 24 in Algorithm 1). The items 'a', 'b', 'c' and 'd' are generated as quasi-periodic-frequent patterns (lines 26 to 28 in Algorithm 1). The items which are not pruned are sorted in descending order of their support values (line 31 in Algorithm 1). The resultant QPF-list is shown in Fig. 2(f). Let PI be the set of potential items that exist in QPF-list.

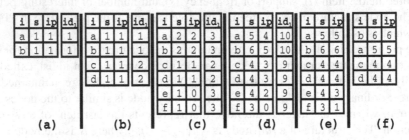

	(a)				(b)				(c)				(d)				(e)			(f)	
i	s	ip	id_l	i	s	ip	id_l	i	s	ip	id_l	i	s	ip	id_l	i	s	ip	i	s	ip
a	1	1	1	a	1	1	1	a	2	2	3	a	5	4	10	a	5	5	b	6	6
b	1	1	1	b	1	1	1	b	2	2	3	b	6	5	10	b	6	6	a	5	5
				c	1	1	2	c	1	1	2	c	4	3	9	c	4	4	c	4	4
				d	1	1	2	d	1	1	2	d	4	3	9	d	4	4	d	4	4
								e	1	0	3	e	4	2	9	e	4	3			
								f	1	0	3	f	3	0	9	f	3	1			

Fig. 2. QPF-list. (a) After scanning first transaction (b) After scanning second transaction (c) After scanning third transaction (d) After scanning entire transactional database (e) Reflecting correct number of interesting periods (f) compact QPF-list containing only potential items.

4.4 Construction of QPF-Tree

With the second database scan, the QPF-tree is constructed in such a way that it only contains nodes for items in QPF-list. We use an example to illustrate the construction of a QPF-tree.

Continuing with the ongoing example, using the FP-tree [7] construction technique, only the items in QPF list take part in QPF-tree construction. For simplicity of figures, we do not show the node traversal points in trees; however, they are maintained in a

Algorithm 1. QPF-list (T: Transactional database, I: set of items, *minsup*: minimum support, *maxperiod*: maximum period, *minpr*: minimum periodic-ratio)

1: Let id_l be a temporary array that explicitly records the *tids* of the last occurring transactions of all items in the QPF-list. Let t_{cur} be the *tid* of current transaction.
2: **for** each transaction $t_{cur} \in T$ **do**
3:     **for** each item i in t_{cur} **do**
4:         **if** t_{cur} is i's first occurrence **then**
5:             $s = 1$, $id_l = t_{cur}$;
6:             $ip = t_{cur} \leq maxperiod?1 : 0$;
7:         **else**
8:             **if** $t_{cur} - id_l \leq maxperiod$ **then**
9:                 $ip++$;
10:            **end if**
11:            $s++$, $id_l = t_{cur}$;
12:        **end if**
13:    **end for**
14: **end for**
15: /*Measuring correct number of interesting periods.*/
16: **for** each item i in QPF-list **do**
17:     **if** $|T| - id_l \leq maxperiod$ **then**
18:         $ip++$;
19:     **end if**
20: **end for**
21: /* Identifying potential items in QPF-list. */
22: **for** each item i in QPF-list **do**
23:     **if** $\left(s < minsup\right) \| (\frac{ip}{(minsup+1)} < minpr)$ **then**
24:         remove i from QPF-list;
25:     **else**
26:         **if** $\frac{ip}{(s+1)} \geq minpr$ **then**
27:             output i as quasi-periodic-frequent item.
28:         **end if**
29:     **end if**
30: **end for**
31: Sort the remaining (potential) items in the QPF-list in descending order of their support values. Let PI be the set of potential items.

fashion like FP-tree does. The tree construction starts with inserting the first transaction $\{a,b\}$ according to the QPF-list order i.e., $\{b,a\}$, as shown in Fig. 3(a). The *tail*-node "$a:1$" carries the *tid* of the transaction. Fig. 3(b) shows the QPF-tree generated in the similar procedure after scanning the second transaction. Fig. 3(c) shows the resultant QPF-tree generated after scanning every transaction in the database.

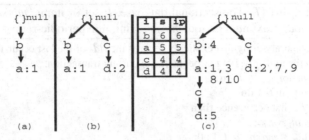

Fig. 3. QPF-tree. (a) After scanning first transaction (b) After scanning second transaction and (c) After scanning complete transactional database.

Based on the QPF-list population technique and the above example, we have the following property and lemmas of a QPF-tree. For each transaction t in T, let $PI(t)$ be the set of potential items in t that exist in QPF-list, i.e., $PI(t) = item(t) \cap PI$, and is called the *potential item projection* of t.

Property 3. A QPF-tree maintains a complete set of all *potential item* projection for each transaction in T only once.

Lemma 3. *Given a transactional database T, a minsup, a maxperiod and a minpr, the complete set of all* potential item *projections of all transactions in T can be derived from the QPF-tree.*

Proof. Based on Property 3, $PI(t)$ of each transaction t is mapped to only one path in the tree and any path from the *root* up to a *tail*-node maintains the complete projection for exactly n transactions (where n is the total number of entries in the *tid*-list of the *tail*-node).

Lemma 4. *The size of a QPF-tree (without the root node) on a transactional database T for a minsup, a maxperiod and a minpr is bounded by $\sum_{t \in T} |PI(t)|$.*

Proof. According to the QPF-tree construction process and Lemma 3, each transaction t contributes at best on path of the size $|PI(t)|$ to a QPF-tree. Therefore, the total size contribution of all transactions can be $\sum_{t \in T} |PI(t)|$ at best. However, since there are usually a lot of common prefix patterns among the transactions, the size of a QPF-tree is normally much smaller than $\sum_{t \in T} |PI(t)|$.

One can assume that the structure of a QPF-tree may not be memory efficient, since it explicitly maintains *tids* of each transaction. But, we argue that the QPF-tree achieves the memory efficiency by keeping such transaction information only at the *tail*-nodes and avoiding the support count field at each node. Moreover, keeping the *tid* information in tree can also been found in literature for efficient periodic-frequent pattern mining [16].

Therefore, the highly compact QPF-tree structure maintains the complete information for all periodic-frequent patterns. Once the QPF-tree is constructed, we use an FP-growth-based pattern growth mining technique to discover the complete set of periodic-frequent patterns from it.

4.5 Mining Quasi-Periodic-Frequent Patterns

Even though both of the QPF-tree and FP-tree arrange items in support-descending order, we cannot directly apply FP-growth mining on a QPF tree. The reason is that, QPF-tree does not maintain support count at each node, and it handles the *tid*-lists at *tail*-nodes. Therefore, we devise a pattern growth-based bottom-up mining technique that can handle the additional features of QPF-tree. The basic operations in mining a QPF-tree for quasi-periodic-frequent patterns are (*i*) counting length-1 potential items, (*ii*) constructing the prefix-tree for each potential pattern, and (*iii*) constructing the conditional tree from each prefix-tree. The QPF-list provides the length-1 potential items. Before discussing the prefix-tree construction process we explore the following important property and lemma of a QPF-tree.

Property 4. A tail-node in a QPF-tree maintains the occurrence information for all the nodes in the path (from that *tail*-node to the root) at least in the transactions in its tid-list.

Lemma 5. *Let $B = \{b_1, b_2, \cdots, b_n\}$ be a branch in a QPF-tree where node b_n is the tail-node carrying the tid-list of the path. If the tid-list is pushed-up to node b_{n-1}, then b_{n-1} maintains the occurrence information of the path $B' = \{b_1, b_2, \cdots, b_{n-1}\}$ for the same set of transactions in the tid-list without any loss.*

Proof. Based on Property 4, b_n maintains the occurrence information of the path B' at least in the transactions in its *tid*-list. Therefore, the same *tid*-list at node b_{n-1} exactly maintains the same transaction information for B' without any loss.

Using the feature revealed by the above property and lemma, we proceed to construct each prefix-tree starting from the bottom-most item, say i, of the QPF-list. Only the prefix sub-paths of nodes labeled i in the QPF-tree are accumulated as the prefix-tree for i, say PT_i. Since i is the bottom-most item in the QPF-list, each node labeled i in the QPF-tree must be a *tail*-node. While constructing the PT_i, based on Property 4 we map the *tid*-list of every node of i to all items in the respective path explicitly in a temporary array (one for each item). It facilitates the support and number of interesting periods' calculation for each item in the QPF-list of PT_i. Moreover, to enable the construction of the prefix-tree for the next item in the QPF-list, based on Lemma 5 the *tid*-lists are pushed-up to respective parent nodes in the original QPF-tree and in PT_i as well. All nodes of i in the QPF-tree and i's entry in the QPF-list are deleted thereafter. Fig. 4(a) shows the status of the QPF-tree of Fig. 3(c) after removing the bottom-most item 'd'. Besides, the prefix-tree for 'd', PT_d is shown in Fig. 4(b).

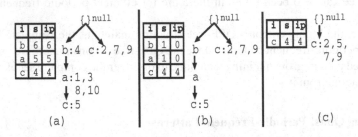

Fig. 4. Prefix-tree and conditional tree construction with QPF-tree. (a) QPF-tree after removing item 'd' (b) Prefix-tree for 'd' and (c) Conditional tree for 'd'.

The conditional tree CT_i for PT_i is constructed by removing the items whose support is less than *minsup* or *mai* value is less than *minpr*. If the deleted node is a *tail*-node, its *tid*-list is pushed-up to its parent node. Fig. 4(c), for instance, shows the conditional tree for 'd', CT_d constructed from the PT_d of Fig. 4(b). The contents of the temporary array for the bottom item 'j' in the QPF-list of CT_i represent T^{ij} (i.e., the set of all *tids* where items i and j are occurring together). Therefore, it is rather simple calculation to compute $S(ij)$, $mai(ij)$ and $periodic\text{-}ratio(ij)$ from T^{ij} by generating P^{ij}. If $S(ij) \geq minsup$ and $periodic\text{-}ratio(ij) \geq minpr$, then the pattern "$ij$" is generated as a quasi-periodic-frequent pattern with support and periodic-ratio values of $S(ij)$ and $periodic\text{-}ratio(ij)$, respectively. The same process of creating prefix-tree and its corresponding conditional tree is repeated for further extensions of "ij". Else, if $mai(ij) \geq minpr$ and $S(ij) \geq minsup$, then the above process is still repeated for further extensions of "ij" even though "ij" is not a quasi-periodic-frequent pattern. The whole process of mining for each item is repeated until $QPF\text{-}list \neq \emptyset$.

For the transactional database in Fig. 1(a), the quasi-periodic-frequent patterns generated at $minsup = 3$, $maxperiod = 2$ and $minpr = 0.8$ are shown in Table 1. The above bottom-up mining technique on support-descending QPF-tree is efficient, because it shrinks the search space dramatically as the mining process progresses.

Table 1. The quasi-periodic-frequent patterns generated for the transactional database shown in Fig. 1(a)

S.No.	Patterns	Support	Periodic-ratio	S. No.	Patterns	Support	Periodic-ratio
1	{a}	6	0.85	4	{d}	4	0.8
2	{b}	5	0.83	5	{c,d}	4	0.8
3	{c}	4	0.8	6	{a,b}	5	0.83

4.6 Relation between Frequent, Periodic-Frequent and Quasi-Periodic-Frequent Patterns

In a transactional database T, let F be the set of frequent patterns generated at $minsup = a$. In T, let PF be the set of periodic-frequent patterns generated at $minsup = a$ and $maxprd = b$. Similarly, let QPF be the set of quasi-periodic-frequent patterns generated at $minsup = a$, $maxperiod = b$ and $minpr = c$. The relationship between these patterns is $PF \subseteq QPF \subseteq F$. If $minpr = 1$, then $PF = QPF$. Thus, the proposed model generalises the existing mode of periodic-frequent patterns.

4.7 Differences between QPF-Tree and PF-Tree

The prefix-structure of QPF-tree is similar to that of PF-tree. However, the differences between these two trees are as follows.

1. The list (i.e., PF-list) of PF-tree captures periodicity (maximum period) of a pattern. The QPF-list of QPF-tree captures the number of interesting periods of a pattern.
2. The periodic-frequent patterns follow *downward closure property*. Therefore, only the periodic-frequent items (or 1-patterns) participate in the construction of PF-tree. The quasi-periodic-frequent patterns do not follow *downward closure property*. Therefore, potential items participate in the construction of QPF-tree.
3. Mining procedure used for discovering periodic-frequent patterns from PF-tree and quasi-periodic-frequent patterns from QPF-tree are completely different. The reason is quasi-periodic-frequent patterns do not satisfy *downward closure property*.

In the next section, we present the experimental results on finding quasi-periodic-frequent patterns from the QPF-tree.

5 Experimental Results

In this section, we first investigate the interestingness of the quasi-periodic-frequent patterns with respect to the periodic-frequent patterns. Next, we investigate the compactness (memory requirements) and execution time of QPF-tree on various datasets with varied *minsup*, *maxperiod* and *minpr* values. Finally, we investigate the scalability of the QPF-tree. All programs are written in C++ and run with Ubuntu on a 2.66 GHz machine with 1 GB memory. The runtime specifies the total execution time, i.e., CPU and I/Os. The experiments are pursued on synthetic ($T10I4D100K$) and real-world datasets (retail, mushroom and kosarak). The $T10I4D100k$ dataset is a sparse dataset containing 1,00,000 transactions and 886 items. The retail dataset is also a sparse dataset containing 88,162 transactions and 16,470 items. The mushroom dataset is a dense dataset containing 8,124 transactions and 119 items. The kosarak dataset is a huge sparse dataset with a large number of distinct items (41,270) and transactions (990,002). These datasets are widely used in frequent pattern mining literature and are available at Frequent Itemset MIning (FIMI) repository [1].

5.1 Experiment 1: Interestingness of Quasi-Periodic-Frequent Patterns

In this experiment, we study the interestingness of quasi-periodic-frequent patterns with reference to the periodic-frequent patterns. Note that at $minpr = 1$, all the quasi-periodic-frequent patterns discovered using the proposed model are periodic-frequent patterns.

The quasi-periodic-frequent patterns generated on the variations of $minsup$, $maxperiod$ and $minpr$ values over several datasets are reported in Table 2. The columns titled "A" and "B" respectively denote the number of quasi-periodic-frequent patterns and maximal length of the quasi-periodic-frequent patterns getting generated at different $minsup$, $maxperiod$ and $minpr$ values.

The data in the table demonstrate the following. First, increase in $minsup$ (keeping other constraints fixed) decreases the number of quasi-periodic-frequent patterns. It is because many items fail to satisfy increased $minsup$ constraint. Second, increase in $minpr$ also decreases the number of quasi-periodic-frequent patterns. The reason is that many patterns have failed to appear periodically for longer time durations. Third, increase in $maxperiod$ increases the number of quasi-periodic-frequent patterns. It is because of the increased interval range in which a pattern should reappear.

Table 2. Quasi-periodic-frequent patterns generated at different $minsup$, $maxperiod$ and $minpr$ values

Database	$minsup$	$maxperiod_1 = 0.1\%$						$maxperiod_2 = 0.5\%$					
		$minpr$=0.5		$minpr$=0.75		$minpr$=1		$minpr$=0.5		$minpr$=0.75		$minpr$=1	
		A	B	A	B	A	B	A	B	A	B	A	B
T10I4D100k	0.1%	624	5	272	1	0	0	20748	10	6360	8	229	2
	1.0%	385	3	272	1	0	0	385	3	385	3	229	2
Retail	0.1%	837	5	293	5	4	2	5849	5	2177	5	15	3
	1.0%	159	4	102	4	4	2	159	4	159	4	15	3
Mushroom	10%	574,431	16	570,929	16	15	4	574,431	16	574,431	16	135	6
	20%	53,583	15	53,583	15	15	4	53,583	15	53,583	15	135	6

In Table 2, it can be observed that at $minpr = 1$, the number of quasi-periodic-frequent patterns (or periodic-frequent patterns) getting generated in different datasets (especially in sparse datasets) are very few, and also the maximal length of these patterns is very less. Whereas, by relaxing the $minpr$, we are able to generate more number of quasi-periodic-frequent patterns, and more importantly, the maximal length of the patterns is relatively large. This shows that relaxing the periodic occurrences of a pattern throughout a database, facilitates the user to find interesting knowledge pertaining to the patterns that are "mostly" occurring periodically in the database.

5.2 Experiment 2: Compactness and Execution Time of the QPF-Tree

The memory consumptions of QPF-tree on the variations of $minsup$, $maxperiod$ and $minpr$ values over several datasets are reported in Table 3. It can be observed that similar observations that are drawn from Experiment 1 can also be observed with respect to the memory requirements of QPF-tree. The reason is that memory requirements of QPF-tree depends on the number of quasi-periodic-frequent patterns getting generated. More importantly, it is clear from the Table 3 that the structure of QPF-tree can easily be handled in a memory efficient manner irrespective of the dataset type (dense or sparse) or size (large or small) and threshold values.

Table 3. Memory requirements for the QPF-tree. Memory is measured in *MB*.

Dataset	minsup	$maxperiod_1 = 0.1\%$			$maxperiod_2 = 0.5\%$		
		minpr=0.5	minpr=0.75	minpr=1	minpr=0.5	minpr=0.75	minpr=1
T10I4D100k	0.1%	11.077	10.975	10.828	11.226	11.197	11.179
	1.0%	8.255	7.820	6.656	8.255	8.255	8.255
Retail	0.1%	4.908	4.193	3.542	6.914	6.254	5.578
	1.0%	1.060	0.942	0.789	1.060	1.060	1.051
Mushroom	10%	0.250	0.250	0.235	0.250	0.250	0.250
	20%	0.131	0.131	0.122	0.131	0.131	0.127

The runtime taken by QPF-growth for generating quasi-periodic-frequent patterns at different *minsup*, *maxperiod* and *minpr* values on various datasets are shown in Table 4. The runtime encompasses all phases of QPF-list and QPF-tree constructions, and the corresponding mining operation. The runtime taken for generating quasi-periodic-frequent patterns depends upon the number of quasi-periodic-frequent patterns getting generated. Therefore, similar observations that were drawn in Experiment 1 can also be drawn from Table 4.

Table 4. Runtime requirements for the QPF-tree. Runtime is measured in seconds.

Dataset	minsup	$maxperiod_1 = 0.1\%$			$maxperiod_2 = 0.5\%$		
		minpr=0.5	minpr=0.75	minpr=1	minpr=0.5	minpr=0.75	minpr=1
T10I4D100k	0.1%	120.533	124.766	128.057	105.951	110.085	113.957
	1.0%	109.257	102.138	83.009	112.400	107.000	112.604
Retail	0.1%	43.780	35.982	30.527	67.782	58.759	50.289
	1.0%	15.441	15.182	14.352	15.780	15.909	15.894
Mushroom	10%	20.580	20.180	16.510	20.320	20.270	18.860
	20%	2.920	2.790	2.780	2.930	3.000	2.740

5.3   Experiment 3: Scalability of QPF-Tree

We study the scalability of our QPF-tree on execution time and required memory by varying the number of transactions in database. We use real *kosarak* dataset for the scalability experiment, since it is a huge sparse dataset. We divided the dataset into five portions of 0.2 million transactions in each part. Then we investigated the performance of QPF-tree after accumulating each portion with previous parts and performing quasi-periodic-frequent pattern mining each time. We fix the $minsup = 2\%$, $maxperiod = 50\%$ and $minpr = 0.75$ for each experiment. The experimental results are shown in Fig. 5. The time and memory in y-axes of the left and right graphs in Fig. 5 respectively specify the total execution time and required memory with the increase of database size. It is clear from the graphs that as the database size increases, overall tree construction and mining time, and memory requirement increases. However, QPF-tree shows stable performance of about linear increase in runtime and memory consumption with respect to the database size. Therefore, it can be observed from the scalability test that QPF-tree can mine quasi-periodic-frequent patterns over large datasets and distinct items with considerable amount of runtime and memory.

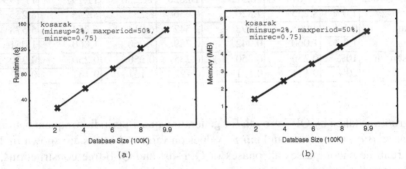

Fig. 5. Scalability of QPF-tree. (a) Runtime and (b) Memory.

6   Conclusions

In this paper, we have exploited the notion "patterns that are mostly appearing periodically in a database are interesting," and introduced a new class of user-interest-based frequent patterns, called quasi-periodic-frequent patterns. This paper proposes a model which facilitates the user to specify the minimum proportion in which a pattern should appear periodically throughout the database. An efficient pattern-growth approach (i.e., QPF-growth) was proposed to mine these patterns. The quasi-periodic-frequent patterns do not satisfy the *downward closure property*. Three pruning techniques have been proposed to reduce the computational cost of mining the patterns. The experimental results from synthetic and real-world databases demonstrate that quasi-periodic patterns can provide useful information. Furthermore, the results have also demonstrated that our QPF-growth can be time and memory efficient during mining the quasi-periodic-frequent patterns, and is highly scalable in terms of runtime and memory consumption.

References

1. Frequent itemset mining repository, http://fimi.ua.ac.be/data/
2. Agrawal, R., Imieliński, T., Swami, A.: Mining association rules between sets of items in large databases. In: SIGMOD, pp. 207–216 (1993)

3. Antunes, C.M., Oliveira, A.L.: Temporal data mining: An overview. In: Workshop on Temporal Data Mining, KDD (2001)
4. Aref, W.G., Elfeky, M.G., Elmagarmid, A.K.: Incremental, online, and merge mining of partial periodic patterns in time-series databases. IEEE TKDE 16(3), 332–342 (2004)
5. Cheng, J., Ke, Y., Ng, W.: A survey on algorithms for mining frequent itemsets over data streams. Knowledge and Information Systems 16(1), 1–27 (2008)
6. Han, J., Dong, G., Yin, Y.: Efficient mining of partial periodic patterns in time series database. In: ICDE, pp. 106–115 (1999)
7. Han, J., Pei, J., Yin, Y., Mao, R.: Mining frequent patterns without candidate generation: A frequent-pattern tree approach. Data Min. Knowl. Discov. 8(1), 53–87 (2004)
8. Uday Kiran, R., Krishna Reddy, P.: Towards efficient mining of periodic-frequent patterns in transactional databases. In: Bringas, P.G., Hameurlain, A., Quirchmayr, G. (eds.) DEXA 2010, Part II. LNCS, vol. 6262, pp. 194–208. Springer, Heidelberg (2010)
9. Kiran, R.U., Reddy, P.K.: An alternative interestingness measure for mining periodic-frequent patterns. In: Yu, J.X., Kim, M.H., Unland, R. (eds.) DASFAA 2011, Part I. LNCS, vol. 6587, pp. 183–192. Springer, Heidelberg (2011)
10. Li, C., Yang, Q., Wang, J., Li, M.: Efficient mining of gap-constrained subsequences and its various applications. ACM Trans. Knowl. Discov. Data 6(1), 2:1–2:39 (2012)
11. Li, Z., Ding, B., Han, J., Kays, R., Nye, P.: Mining periodic behaviors for moving objects. In: KDD 2010, pp. 1099–1108 (2010)
12. Ma, S., Hellerstein, J.: Mining partially periodic event patterns with unknown periods. In: ICDE, pp. 205–214 (2001)
13. Özden, B., Ramaswamy, S., Silberschatz, A.: Cyclic association rules. In: ICDE, pp. 412–421. IEEE Computer Society, Washington, DC (1998)
14. Rashid, M. M., Karim, M. R., Jeong, B.-S., Choi, H.-J.: Efficient mining regularly frequent patterns in transactional databases. In: Lee, S.-G., Peng, Z., Zhou, X., Moon, Y.-S., Unland, R., Yoo, J. (eds.) DASFAA 2012, Part I. LNCS, vol. 7238, pp. 258–271. Springer, Heidelberg (2012)
15. Sheng, C., Hsu, W., Lee, M.L.: Mining dense periodic patterns in time series data. In: ICDE, Washington, DC, USA, pp. 115–117 (2006)
16. Tanbeer, S.K., Ahmed, C.F., Jeong, B.-S., Lee, Y.-K.: Discovering periodic-frequent patterns in transactional databases. In: Theeramunkong, T., Kijsirikul, B., Cercone, N., Ho, T.-B. (eds.) PAKDD 2009. LNCS, vol. 5476, pp. 242–253. Springer, Heidelberg (2009)
17. Yang, J., Wang, W., Yu, P.S.: Mining asynchronous periodic patterns in time series data. IEEE Trans. on Knowl. and Data Eng. 15(3), 613–628 (2003)
18. Yang, R., Wang, W., Yu, P.: Infominer+: mining partial periodic patterns with gap penalties. In: ICDM, pp. 725–728 (2002)
19. Zhang, M., Kao, B., Cheung, D.W., Yip, K.Y.: Mining periodic patterns with gap requirement from sequences. ACM Trans. Knowl. Discov. Data 1(2) (August 2007)

Challenges and Approaches for Large Graph Analysis Using Map/Reduce Paradigm

Soumyava Das and Sharma Chakravarthy

University of Texas at Arlington
soumyava.das@mavs.uta.edu sharma@cse.uta.edu

Abstract. Analysis of big graphs has become possible due to the convergence of many technologies such as server farms, new paradigms for processing data in massively parallel ways (e.g., Map/Reduce, Bulk Synchronous Parallelization), as well as the ability to process unstructured data. This has allowed one to solve problems that were not possible (or extremely time consuming) earlier. Many algorithms are being mapped to new paradigms to deal with larger versions with a meaningful response time.

This paper analyses a few related problems in the context of graph analysis. Our goal is to analyse the challenges of adapting/extending algorithms for graph analysis (graph mining, graph matching and graph search) to exploit the Map/Reduce paradigm for very large graph analysis in order to achieve scalability. We intend to explore alternative paradigms that may be better suited for a class of applications.

As an example, finding interesting and repetitive patterns from graphs forms the crux of graph mining. For processing large social network and other graphs, extant main memory and disk-based) approaches are not appropriate. Hence, it is imperative that we explore massively parallel paradigms for their processing to achieve scalability. This is also true for other problems such as in-exact or approximate match of graphs, and answering queries on large data sets represented as graphs (e.g. Freebase). In this paper, we identify the challenges of harmonizing the existing approaches with Map/Reduce and our preliminary approaches to solve these problems using the new paradigm.

1 Introduction

As a general data structure, graphs have become increasingly important in modelling sophisticated structures and their interactions, with broad applications including chemical informatics, bio-informatics, computer vision, video indexing, text retrieval and web analysis. With the rapid development of social networks and web, the graphs have grown too large to be fit in one single machine. The onus is now on the developer/system to partition the graph across several machines to achieve satisfying computation time and formidable scalability. Hence a natural question that follows is that, how the existing graph algorithms will work on these partitioned graphs. It will also be interesting to see how much

V. Bhatnagar and S. Srinivasa (Eds.): BDA 2013, LNCS 8302, pp. 116–132, 2013.

speedup is achieved while running the program on multiple machines instead of a single machine.

Mining interesting and repetitive patterns in a graph lays the stage for identifying concepts in the graph which are not apparent at the onset. Graph mining for concept identification deals with two basic aspects: systematic expansion from smaller structures to larger structures and aggregating similar (exact and inexact) structures to form a concept. Grouping same substructures form an integral part of mining where the main goal is to match one substructure with another. As a result, graph matching or comparing similarity of graphs became considerably important. As the matching methods are established, matching a graph in a set of graphs or a single graph (graph querying) became important. As graph mining, matching and querying are related together solving one of them would set the stage for investigating the others.

Graph mining is much different from traditional mining approaches. Traditional mining approaches like association rule mining, decision trees mine transactional data and do not make use of any structural information present in the data set. On the other hand, graph-based techniques are used for mining data that are structural in nature. A few examples of such data are phone call networks, complex protein, entity relationship graphs (e.g. Freebase), VLSI circuits, social networks etc. Traditional mining approaches on these datasets would result in loss of structural property. Existing graph mining methods will also face scalability issues while dealing with such huge graphs. Map/Reduce, a distributed programming framework with all implemented features like auto restart, load distribution and fault tolerance was introduced by Google [11] which can process huge amounts of data in a massively parallel way using simple commodity machines. The main advantage of Map/Reduce with existing distributed paradigms is that Map/Reduce can tackle graphs of massive sizes and achieve high scalability without the user worrying for auto restart or fault tolerance. In our current work we investigate in detail how existing graph mining algorithm (SUBDUE) can be made scalable for larger graphs and present a parallel version of substructure enumeration problem using Map/Reduce.

Graph matching on the other hand deals with the challenge of matching two or more graphs. When complex structured objects are represented as graphs the problem of measuring object similarity transforms into the problem of matching one graph with another. Two graphs are exactly similar or isomorphic if there exists a one to one mapping between each node and edge of one graph with that of the other. But as similarity of objects assumes a score between 0 and 1, graph matching should also follow such a norm resulting in inexact match or approximate matching of two graphs. Graph matching finds use in graph search and query where the goal is to match the query graph in the target graph both in an exact and inexact manner. Our aim in this work is to analyse the challenges of graph matching using Map/Reduce, analyse their performance against the older ones thus laying the platform for using graph matching in querying of graphs in a plethora of application domains.

Graph matching lays the stage for graph search and query where the goal is to match the query graph in the main graph(s). The rapid proliferation of structured networks (entity-relationship graphs, social networks etc.) has brought in the need of for supporting efficient querying and mining methods over structured graph data. Graphs model structured data and to query on graphs the query should also be structured. Thus came the need of representing query as graphs which are also referred to as graph queries. Querying graph databases mainly falls into two categories. The first category considers querying on a large number of small graphs or transactional graph database. Here an example query can be to find all the chemical compounds which contain the specified drug group. The second category of queries are on one single big graph as in bibliographical and social networks. For example an interesting query would be to find all Stanford graduates who founded a company in Silicon Valley. The graph queries are different from each other and hence different partitioning strategies and methodologies need to be used to handle these two type of queries. If the query does not find an exact result it is meaningful to return some results which are similar to input query graph thus involving the heavy use of graph matching in query answering in large networks.

In order to analyse big graphs and obtain results within a meaningful response time, the input data needs to be processed using a paradigm that can leverage partitioning and distribution of data on a large scale. Map/Reduce is one such paradigm that may be beneficially applied to partition, compute, merge, and synchronize computations to obtain meaningful results.. The basic goal of Map/Reduce has been to use commodity machines for processing very large amounts of raw data without having to load them into a database or setup customized ways for processing each data set. Hadoop [1] is an open source implementation of Map/Reduce which provides a powerful programming framework to work with such large data sets. The reason behind choosing Map/Reduce over other distributed paradigms is its inbuilt features of auto restart, load balancing and fault tolerance. Our goal is to figure out how to do graph mining, substructure comparison, and matching queries expressed as graphs in a forest of graphs for both exact and inexact matching using the Map/Reduce paradigm. We also need to address how to do this in such a way that we achieve scalability with no loss of information, and appropriate speedup based on the data set size and the number of processors used.

RoadMap: Section 1 introduces the problems we are trying to analyse. Section 2 summarizes the two paradigms which are of interest to us. Sections 3, 4 and 5 highlights the challenges and proposed approaches in the respective areas of graph mining, graph matching and querying in graphs. Section 6 concludes the paper.

2   Distributed Paradigms

Since its introduction distributed computing acquired great popularity in a plethora of application domains. The domains include telecommunication networks, wireless sensor networks, real time control systems as well as scientific computing. The initial methods for distributed processing started with shared memory architecture between different nodes. However with shared memory the programmer has to choose which node should run which program. Message Passing Interface (MPI) was developed which removed this aspect of shared memory models by running the same program on all nodes. However in MPI each node can send message to any other node thus making it useful to address communication-intensive task. Map/Reduce on the other hand is more suited for data-intensive task. Map/Reduce is easier to learn while MPI is complex with lots of functions. Moreover Map/Reduce provides better fault tolerance that is when one node fails the task can be started at another node unlike MPI where failure of one node can terminate the entire system. Grid computing followed soon where a grid was a set of same class of computers connected from multiple locations to reach a common goal. Grid computing provides a platform where resources can be dynamically linked and used to support the execution of applications that require significant amounts of computer resources. However in grid computing, the results of all processes are sent to all nodes in the grid and then the final result is assessed. Grid computing requires small servers, fast connections between the servers and the quality tools, software and skilled technicians to manage the grid and maximize the productivity. When all these components are put together, it is obvious that this technology is costly. Map/Reduce offers a simple, elegant and cheaper solution by using simple commodity machines.

Map/Reduce: Users in earlier distributed processing methods have to implement fault tolerance, auto restart and also the communication between different machines. Map/Reduce system architecture takes care of all of these leaving the user just to implement two functions map and reduce. Map/Reduce is a paradigm for partitioning data and processing each partition separately using commodity machines. The input data is stored as $< key; value >$ pairs and the computation proceeds in rounds (or iterations). Each round is split into three consecutive phases: map, shuffle, and reduce. In the map phase, each line of the partition, represented as a $< key; value >$ pair, is evaluated by a processor (termed a mapper) producing some output $< key; value >$ pair(s). The next phase shuffle groups all the intermediate mapper outputs across all the mappers by the *key*. Then it partitions the resulting set of $< key; value >$ pairs based on the number of reducers and each partition (may contain several $< key; value >$ pairs) is sent to a distinct reducer.This paradigm has been automated in the cloud framework to allocate mappers and reducers and also to manage fault-tolerance in the presence of resource failures. In addition, HDFS (Hadoop Distributed File System) and local file support (LFS) are provided to store data globally and locally, respectively. The number of partitions of data (and hence the number of mappers/reducers) is controlled by the problem solver (or will be partitioned using some default size by the system). The number of iterations is also determined by the algorithm.

Pregel: The Pregel [18], introduced by Google for distributed processing of large graphs is inspired by the Bulk Synchronous Parallel (BSP) model [30]. The computations of Pregel consists of a sequence of iterations known as *superstep*. During a superstep the framework invokes a function for each of the vertex of the graph, conceptually in parallel. The function specifies the action of a vertex V at a certain superstep S. In the superstep S, the vertex V can read the data sent by other vertices to it in superstep S-1 and can do local computations and send values to other vertices for superstep S+1. Messages are usually sent along the outgoing edges but there is provision to send values to a vertex with a known identifier even if it is not directly connected to the vertex V. The vertex-centric approach is reminiscent of Map/Reduce where each vertex functions independently of one another. Algorithm termination or convergence is based on each vertex *voting to halt*. Generally the algorithm terminates when all the vertices vote to halt deciding that the algorithm has reached convergence. The main advantage of Pregel over Map/Reduce is guaranteeing convergence in iterative algorithms.

In the next sections we introduce our areas of interest and analyse the challenges and solutions of developing parallel algorithms in these areas.

3   Graph Mining

Graph mining is the process of finding out all interesting and repetitive patterns from a set of graphs or a big single graph. Graph mining is important in detecting presence of abnormalities in a graph because abnormalities will tend to deviate from the regular patterns. Graph Mining also finds use in graph compression as frequent graph patterns represent regularities in the graph and these patterns can be grouped together while compressing the graph. These patterns are smaller subgraphs of the dataset and so graph mining is also referred to as substructure mining. Most of the real world graphs have structural relationships as they not only have user information (corresponding to nodes and node labels), but also relationships (edges and edge labels) among various user objects. The number of unique node and edge labels are much smaller than the number of nodes and edges in the graph making them different from the ones usually studied in literature. Several algorithms exist for substructure mining. Most of these algorithms work in a bottom up manner by generating k-edge substructure using (k-1)-edge substructures.

3.1   Existing Work

For growing a smaller edge substructure into a bigger edge substructure, the algorithms choose to use either vertex growth or edge growth approaches [17]. In the vertex growth scenario the graph is expanded by adding one more vertex to the existing graph while in edge growth one edge is added to any of the vertex in the existing substructure. Edge growth is undoubtedly better than its competitor because it can handle multiple edges in the graph. Graph mining includes a special class of algorithms which finds interesting and repetitive patterns in input graph(s). However little work exists for implementation of these algorithms in the

Map/Reduce paradigm for faster processing and scalability. As the aim of graph mining algorithms is to find interesting and repetitive patterns in a graph, the crux of these algorithms is the isomorphism checking phase or similar subgraph detection. The iterative nature of the basic algorithm has led to the development of three major approaches to deal with graph mining.

Main Memory Approach: Main memory methods were used when the graph was relatively small and could fit into main memory. Earlier approaches and algorithms for graph mining were developed by AI researchers whose goal was identifying important concepts in a structured data set. These algorithms loaded a complete representation of the graph (either in the form of an adjacency list or a matrix) into memory for efficiency reasons. SUBDUE [13] laid the foundation of graph mining where substructures were evaluated using a metric called Minimum Description Length (MDL) principle. The substructure discovery algorithm used by SUBDUE is a computationally-constrained beam search. The algorithm begins with the substructure matching a single vertex in the graph. Each iteration through the algorithm selects the best substructures and expands the instances of these substructures by one neighbouring edge in all possible ways. The algorithm retains the best substructures in a list, which is returned when either all possible substructures have been considered or the total amount of computation exceeds a given limit. The evaluation of each substructure is guided by the minimum description length principle and background knowledge rules provided by the user. Apriori Graph Mining (AGM) [14] followed SUB-DUE and was the first to propose apriori-type algorithm for graph mining. It helped to detect frequent induced subgraph for a given support. Frequent Subgraph Discovery (FSG) [12] aimed at discovering interesting subgraph(s) that occurred frequently over the entire set of graphs. FSG was the first to use the concept of canonical labelling for graphs and worked with the intuition that two similar graphs should have the same canonical labelling. The gSpan algorithm [31] addressed the two additional costs of FSG and AGM algorithms: 1) Costly Sub Graph isomorphism test and 2) Costly candidate generation. gSpan involved a Depth First Search (DFS) technique as opposed to the Breadth First Search (BFS) used inherently in Apiori like algorithms. All the main memory algorithms need to keep a data structure in memory and so is not scalable for graphs of bigger sizes.

Disk-Based Approaches: As more and larger applications with structural information became common place, disk-based graph mining techniques [5,10,23] were developed to overcome the problem of storing the entire data set in main memory. A section of data was kept in memory and the rest was on the disk. As data reside on disk, the question of using sequential or random accessing gained prominence. Sequential data access from disk is much faster than random access when the elements to be accessed have different sizes. If the elements have the same size, then random access is better. Moreover an indexing technique helps in accessing when the elements to be accessed are of variable sizes. As a big graph can be indexed by the frequent substructures present in it, as in gIndex [32] and GBLENDER [15], mining frequent patterns from graphs became even more meaningful.

Database-Oriented Approach: Disk-based algorithms solve the problem of keeping portions of the graph in memory for processing. However, these algorithms needs to marshal data between external storage and main memory buffer and this has to be coded into the algorithm. The performance can be very sensitive to optimal transfer of data between disk and memory, as well as buffer size, buffer management, and hit ratios. An alternative approach is to make use of efficient buffer management and query optimization – already mature in the context of a Database Management System (DBMS) – by mapping these graph mining algorithms to SQL and store data in a DBMS (Database Management System). However SQL came with its bag of advantages and disadvantages. Since joins were used for substructure expansion, duplicate generation could not be avoided. Therefore duplicate elimination before isomorphism testing became an important task for these algorithms. HDB-SUBDUE [22] is the last modification over SUBDUE which handles directed edges, exact graph match and easily handles cycle and loops. Unlike its predecessors, HDB-SUBDUE can handle graphs with millions of edges and vertices. However a huge number of joins and transformations for the duplicate elimination phase increased the runtime of the algorithms. Graphs in modern days span multiple Giga-bytes, and heading to Tera- and Peta-bytes of data. So efforts have been made to identify several other methodologies for disk based graph mining. A promising tool is parallelism, and specially Map/Reduce. Map/Reduce has been used in graph mining in many applications. Map/Reduce came with its own constraints but guaranteed better scalability than the existing disk based approaches.

3.2   Challenges and Proposed Approaches

The first challenge was the change of file system and adaptation of the algorithm to the new file system. Map/Reduce and its open source version (Hadoop) comes with a distributed file system for storing data globally while each node in the cluster of computers has its own local file system for local storage. Sequential access works best with HDFS as the HDFS sends the client the required data and the data following it. Access from HDFS to any client happens by using a *heartbeat communication* which is a type of TCP connection. So in sequential access only one TCP connection is opened as the HDFS always sends the next connected data through the same connection. But a random access in HDFS is also treated as a sequential access. The random accessing is a 3 way method where the client first finds the server and opens a TCP connection followed by the server sending the data packets and finally terminating the connection. Thus each random access requires a new TCP connection. So random access from HDFS is not a suitable method to access any information from HDFS as frequent TCP initializations would increase the overhead severely. One solution to this problem is formation of intelligent partitioning schemes where each node will work independently on a partition. We have dealt with this problem by introducing a range partitioning scheme on the adjacency list. The adjacency list is sorted on the vertex identifiers and then partitioned into k parts with equal number of vertices in each partition. The expansion phase sends the (k-1)-edge substructure to the appropriate partition for expansion accordingly.

The second challenge faced was to decide how many computing nodes in the cluster (mapper/reducer) would be required by the program to achieve a desirable speed-up. The number of mappers in the system depends either on the number of partitions of the input data or on the output of the reducer from another Map/Reduce process. As we are investigating the iterative method of substructure mining, the initial mapper input is from partitioned 1-edge substructures while the subsequent inputs are from outputs of reducer of previous iteration. The number of reducers is usually determined by the number of keys formed during the shuffle stage. For uniform load distribution, the number of reducers are factors of the number of keys generated during the shuffle phase of the input.

The third and the most important challenge was to translate the serial algorithm into meaningful key/value pairs for Map/Reduce. As the input 1-edge list has been partitioned into parts and mappers and reducers work independently of one another, different substructures can grow to the same substructure in the expansion phase. These duplicates can arise while expanding two substructures in the same partition or two substructures in different partitions. As the reducer groups by the keys, one interesting aspect would be to have the same key for similar substructures (duplicates and isomorphic substructures). The duplicates and isomorphic substructures have equal number of nodes and edges and same edge labels, node labels and connectivity. The duplicates will have exactly same vertex identifiers while isomorphs will differ in at least 1 vertex identifier. So a key should be composed of edge labels, node labels and connectivity information to group both isomorphic and duplicate substructures. However there should be a key for each edge in a k-edge substructure as each edge has edge label, source node label, destination node label and connectivity information. Hence, these k key tuples must be arranged in some order that similar substructures have the same ordering. We use the lexicographic ordering scheme where the keys are arranged in an increasing order of edge label, followed by source vertex label, destination vertex label and connectivity information. As the size of keys and values grow across iterations, decreasing number of substructures however do not guarantee decrease in the size of shuffled bytes in each iteration and so the running time of each iteration is not proportional to the number of substructures across iterations.

Summary: We developed mrSUBDUE (Map/Reduce based substructure discovery) to compute best substructures at each iteration. Each iteration is a chained Map/Reduce job where the first job assists in expansion while second job helps in duplicate elimination and isomorphism checking. The initial input was a list of 1-edge substructures and a collection of adjacency list of all vertices. This method used a ranged partitioning scheme on adjacency list to expand a substructure in the proper partition. To bring down each iteration from a chained Map/Reduce job to a single Map/Reduce job we also developed a method where in a single Map/Reduce job the mapper does substructure expansion while the reducer aids in isomorphism detection. Our initial belief was that one Map/Reduce would perform better than a chained Map/reduce job per iteration. However to move from chained Map/Reduce to a single Map/Reduce required a different partitioning strategy. In this approach, 1-edge list was partitioned and each partition was also associated with a partition of adjacency list which held the adjacency list of all the vertices in that edge list partition. In each iteration, the mapper loaded the adjacency list partition to expand, while the reducer modifies the adjacency list by adding adjacency list of newly discovered vertices in the expansion phase. The new adjacency list partition is accepted by the mapper in the next iteration to grow to bigger substructures. The one Map/Reduce method performed at least 18% better than mrSUBDUE for graphs till size of 20,000 vertices and 40,000 edges with 1 mapper and 1 reducer. However as the size of the graphs kept on growing mrSUBDUE started performing better and had a 67% improvement over its counterpart for a graph with 400,000 vertices and 800,000 edges.

4   Graph Matching

The graph matching problems deals with matching two or more graphs. There are two types of matchings in real world. Exact matching is characterised by the notion of edge preservation in the sense that if two nodes in the first graph are connected by an edge then they are mapped to two nodes of another graph linked by an edge as well. A more stringent condition of matching is that the labels of the nodes, the edges and the connectivity should match too. Thus there should exist a one to one correspondence between each node and edge of one graph with that of the other. Such a matching is also referred to as subgraph isomorphism. A slightly weaker notion of matching is homomorphism where one node in first graph can match to more than one node in another graph. Finally one more interesting matching characteristic is where a subgraph of the first graph can be matched with a subgraph of the second. Since such a mapping is not uniquely defined the goal is to find the maximum common subgraph (MCS) of the two graphs. The stringent conditions imposed by exact graph matching needs to be relaxed in some cases where the graphs are subjected to deformation due to several causes: variability of patterns, introduction of noise etc.. Even if there is no deformation finding inexact match can be useful. As seen, the exact match algorithms have an exponential time complexity in the worst case and it is worth exploring if a faster method can give us an approximate match in reasonable time. Graph matching has been a well studied problem in computer science with application in Biology, Computer Vision and chemical compound analysis.

4.1   Existing Work

Many similarity measures have been suggested and studied to explore graph matching. These methods can be categorized into three types:

Edit Distance: Here the similarity of 2 graphs is measured as the cost of transformation of one graph into another. Two graphs are similar if they are isomorphic [24], or one is isomorphic to a subgraph of other(containment) or two subgraphs of these graphs are isomorphic to one another. The graph edit distance is a generalization of the graph isomorphism problem where the goal is to minimize the cost of graph transformation [9]. Generally a notion of cost is associated with all such transformation operations like insertion, deletion and substitution. Moreover metrics like missing edges, graph edit distance are appropriate for chemical compound analysis and biological networks. Social networks however are quite different from these physical networks and are often full of noise. Entity relationship graphs (e.g., Freebase) and social network graphs because of noise are not always interested in exact topological matches. Moreover these graph matching methods do not scale well for big networks. For e.g., the edit distance metric compares each vertex with every other vertex with the graph giving rise to an exponential complexity. However branch and bound approach used by SUBDUE makes the algorithm computationally bound as number of mappings are far less than the original algorithm. Inexact match in SUBDUE is achieved by comparing both same and different sized graph by using a similarity measure. the similarity measure in SUBDUE is the number of mappings needed to make both of them exact by i) change of edge labels ii) addition of nodes and edges iii) deletion of nodes and edges. However as SUBDUE is a main memory algorithm and has a high initialization time for big datasets, the need is to develop graph matching algorithms (for both exact and approximate match) that would scale well as the size of the graphs keep on increasing.

Feature Extraction: Similar graphs have similar features. The mostly studied features have been degree distribution, diameter [8], eigen value [29] and compressibility. These basic graph invariants play an immensely popular role in comparing similarity of 2 graphs. After extracting these features, a similarity measure is applied over the aggregated statistics. These methods scale well, as they map the big graph to several smaller statistics which are much less than the graph size. The results of the similarity are not always intuitive as it is possible to get high similarity between two graphs even though they have different node set size. Investigating these features for e.g, as finding diameter or compressibility in big graphs is not straightforward. So the need is to use parallel processing for faster evaluation of these features in big graphs.

Iterative Methods: The intuition behind iterative methods is that two graphs are similar if their neighbours of node set are similar. In each iteration nodes exchange their values and the process ends when the convergence is achieved.

Database schema matching [19] attempts to solve this problem by finding similarity between two nodes in a graph. The process starts by creating a pairwise connectivity graph and later an induced propagation graph on which the similarity scores are propagated till convergence. Instead of looking only at the similarity score of nodes, Zager and Verghese [33] proposed an algorithm which considers both node and edge similarity. The method compare all pair nodes and all pair edges to find matching between two graphs. Bayati et. al. [7] introduced two message passing approximate algorithms to find similarity between two graphs. The aim is to create a bipartite graph between all nodes of two graphs and find the set of edges in the bipartite graph which do not share any vertex. Creating the pairwise connectivity graph, all pair edges requires combing two graphs of more and less equal sizes and so for massive graphs creating such a pairwise connectivity graph would be computationally expensive for main memory approaches. For faster processing and scalability distributed approaches needs to be explored to evaluate these algorithms.

As the size of the graphs keep on growing, distributed methods for similarity evaluation in graphs also becomes important. Comparing canonical labelling may be considered as one such approach of similarity check. Our research interest lies in faster evaluation of similarities in large graphs. We consider canonical label as one interesting measure for similarity check because of its property that the canonical labelling of a graph is always lexicographically sorted. Thus addition, removal or substitution would appear as a small change in the canonical label of the graph.

4.2   Challenges and Proposed Approaches

The first challenge faced while generating distributed graph matching algorithms was how the existing graph matching algorithms can be made parallel. Our intuition behind this method is that if the graph is divided into smaller partitions, similarity of the smaller partitions would contribute towards similarity of the bigger graph. The main challenge here is to partition both the graphs into equal number of partitions and then decipher a one to one mapping between the partitions to maximize the similarity. As we are working with labelled graphs, the labels play an important part in partitioning the graphs. If the edges with similar node and edge labels and connectivity are grouped together in a single partition, then similar partitions from both graphs can be compared to find number of edge mismatches between the graphs which can then be used as a similarity measure.

The next challenge faced for converting iterative matching algorithms into their parallel counterparts is to create a parallel pairwise connectivity graph and decide when the convergence is reached. In the Map/Reduce framework creating a parallel pairwise connectivity graph (PCG) would not be straight forward. If the PCG can be created, convergence of the propagation values in the PCG is difficult to achieve. To parallelize such a method where convergence needs to be achieved, a solution is to seek other paradigms apart from Map/Reduce which offers such a convergence criteria. One such existing paradigm is Google's

Pregel, it's open source version being Apache Hama [2]. Pregel works using the Bulk Synchronous Parallelization (BSP) model where each vertex in the graph can exchange information with the other by message passing and convergence is achieved when all vertices stop sending messages.

Another solution to graph matching problems is to use canonical labels or DFS lexicographic ordering to find a difference between ordering of the two big graphs. As we are considering graphs of same size initially in our approach, the graphs can be split identically into equal number of partitions and the canonical label or the dfs lexicographic ordering of each partition can be compared for similarity. We can also use mrSUBDUE to hierarchically compress both the graphs and compare the compressed graphs for similarity. The intuition behind it will be that similar graphs would contain similar patterns embedded in them.

5   Querying in Large Graphs

The graph querying problem deals with finding all instances of a given sample graph in a big graph or a set of graphs(forest). Big graph databases like Freebase [3] are structured collection of data harvested across different sources and spanning multiple domains like music, sports etc. The graph is laid out as a set of RDF triplets where each triplet is of form $< subject, predicate, relation >$. Such a triplet preserves the direction between the source label and the destination label and the relation being the edge label. The same subject, predicate and relation can appear in more than one edges. The aim of the user is to enter a query and get meaningful results. For example a meaningful query would be "Find the companies in silicon valley founded by Stanford graduates" or "Find the player who won a gold medal in 2012 Olympics and a silver medal in 2008 Olympics in swimming ". We focus on a special type of query called graph query where the keyword query can be represented as a connected small graph. Generally queries made in such databases are of small size spanning over 5-10 nodes. If the queries find a one to one correspondence with a subgraph in the datagraph we call it an exact match. In cases where exact match is not possible it is necessary to finetune the system to give some approximate matches to user which may be useful. So the graph querying problem can be deemed as a subset of the graph matching problem. There are two types of matching, exact match and inexact match or approximate matching. There are three different types of approaches which handle graph matching.

5.1   Existing Work

Exact query answering has been one of the most popular problems in big data analytics. Several techniques have been employed to fetch exact matches of an input graph in the main graph. Firstly Indexing techniques have been used extensively for graph matching. Edge indexing techniques are used for SPARQL queries on RDF data [6,21]. SPARQL queries are decomposed into a set of edges and multi-way joins are engaged to find the results. To avoid excessive joins another approach was to mine and index a graph by all frequent subgraphs [32]. However this method is not good for queries which are infrequent. Moreover for a big graph with millions and billions of nodes the time to create one such index is huge [25]. Therefore some graph matching applications are done without any indexing. Trinity Cloud by Microsoft uses an in-memory graph exploration technique for graph matching [25]. Thirdly, some methods like GBLENDER [16] makes use of query creation latency to generate intermediate results and then guarantee a better system response time. Map/Reduce has been used in graph matching problems for triangle enumeration [26]. Recently Afrati [4] used a single round of Map/Reduce to enumerate all instances of a given sample graph in a main graph. Their intuition is to generate a set of conjunctive queries which consist of ordered relations in the input query graph. The basic intuition behind all of these approaches to find a query decomposition in a particular order which would make querying easier by pruning the search space effectively.

Approximate query answering has also been a well cultivated problem. There has been different definitions for inexact match in the research world. GRAY [28] finds inexact matches that preserves the shape of the input query. TALE [27] on the other hand considers top k approximate subgraph matches based on number of missing edges. SIGMA [20] is a set cover based method that defines inexact match as structures differing from input structure by at most r edge deletions. Most of these definitions for inexactness have been based on the particular application domains. With the graphs getting bigger and the exact match problem in graph studied critically, for approximate matching of input graphs two approaches can be used: 1) Generate similar queries from input graph query and make exact matching on them. Here we shall need a measure to identify how close is the transformed graph to the input query graph 2) Query decomposition where the input query graph is decomposed into smaller components and then smaller components are grown both in exact and inexact manner to find matches. Here we shall also need a metric to measure similarity of the output graphs to the main input query graph. Our aim in this thesis is to design both sequential and parallel query answering algorithms for large graphs.

5.2   Challenges and Proposed Approaches

The first approach to solve the graph query is by directly using query matching. The incoming query needs to be matched in the main graph to find the query result. Another technique to solve the process of graph query is by indexing. We can generate and index all the frequent substructures generated in our mining phase. However this faces the problem to address the infrequent queries. Moreover mining frequent substructures from a graph is time consuming and for huge graphs building and maintaining such an index may be an infeasible practice. Another indexing method would be to index local k-hop neighbourhood information with each vertex. As the query is received starting from a node the query results can be explored following the k-hop neighbour information with each query vertex in the main graph. However in this method choosing value of k is difficult. A value of $k = 1$ would give exact matches only as we are looking for exact match of each query edge in the main graph. Keeping higher values of k will give different inexact matches depending on the exploration technique used.

Another method that can be used is to do graph mining initially and create some intermediate results that helps in expansion. For example if we decompose the query into a set of 1-edge substructures, we can find out which information we need from which partitions to expand it. If our goal is to search a query in a graph database or a set of graphs, the graphs can be partitioned based on their edge and node labels into different partitions. Each partition will contain a set of graphs with similar node labels and edge labels. So from the nodes and labels of the input graph we shall know in which partition to find the query graph. However if our goal us to match a query in one large graph, the graph may be partitioned into smaller parts depending on the node and edge labels. However to prevent loss of information some edges needs to be replicated across partitions. Thus each partition will contain a set of edges of same labels or nodes with same labels and the partitions will be used beneficially for query expansion.

Most of the proposed approaches for graph search can follow two techniques, join and exploration. In case of join the main aim is to decompose the query into a set of smaller queries, get their results and aggregate them produce the final results. In query exploration techniques, we start with a node in the query graph and start expanding it and getting partial results at each stage to grow into the results for the entire query. A comparison of joining and query exploration would suggest query exploration to be better suited for graph querying. This is because in case of graph exploration, no index structure is required, costly joins are avoided and less number of partial results are generated in each step of the query exploration. However all queries cannot be answered using exploration technique as the presence of cycles and loops in the query requires joining. Moreover in completely connected graphs the query exploration would result in a huge search space while with proper hash and merge strategies the joining can proceed in batches. So comparing the pros and cons of both joins and explorations suggest that a combination of them would be the best strategy for distributed querying in graphs. The query can be decomposed into several portions by minimizing the number of joins and then each portion may be explored independently in a parallel fashion to get partial results. The partial results then can be joined to get the final results.

One major aspect of graph search is to return approximate results if the exact match is not found. To deal with inexact match, the definition of inexactness is also a challenge. Inexactness has been defined in literature as missing node, missing edges and addition of nodes and edges. So the query exploration should take care of the inexactness during the exploration phase so as to generate approximate results. An alternative to this can be generation of approximate queries apriori from the initial query. The underlying schema graph in entity relational graphs is known beforehand and using this schema graph approximate queries can be created from the given query. One major challenge is to decide how similar is the approximate query is to the initial query as the more similar the approximate query is with the initial query more better will be the quality of output results. The best approximate queries thus found can be executed in parallel to find out a set of results for the input query.

6   Conclusion

The aim of our work is to investigate the challenges of graph analysis when moving from existing methodologies to a different framework (e.g. Map/Reduce). We have identified three connected graph analysis problems (graph mining, graph matching and graph query) for big graphs and have explored the challenges encountered while scaling current methods to big graphs. The critical analysis introduces us to the difficulties faced by existing methods to work on massive graph dataset and lay the platform to evaluate them using the Map/Reduce paradigm. In our future work we shall investigate if Map/Reduce is the best paradigm to analyze graphs or the same class or problems can be solved better by other paradigms using Message Passing or Bulk Synchronous Processing. Extensive experimental analysis would be done to analyse the performance of the graph analysis methods on massive graphs using multiple machines to investigate speed-up and scalability of our proposed approaches.

References

1. http://hadoop.apache.org/
2. http://hama.apache.org/
3. http://www.freebase.com
4. Afrati, F.N., Fotakis, D., Ullman, J.D.: Enumerating subgraph instances using map-reduce. Technical report, Stanford University (December 2011)
5. Alexaki, S., Christophides, V., Karvounarakis, G., Plexousakis, D.: On Storing Voluminous RDF Descriptions: The Case of Web Portal Catalogs. In: International Workshop on the Web and Databases, pp. 43–48 (2001)
6. Atre, M., Chaoji, V., Zaki, M.J., Hendler, J.A.: Matrix "Bit" loaded: a scalable lightweight join query processor for RDF data. In: World Wide Web Conference Series, pp. 41–50 (2010)
7. Bayati, M., Gleich, D.F., Saberi, A., Wang, Y.: Message-passing algorithms for sparse network alignment, vol. 7, p. 3 (2013)
8. Bollobs, B., Chung, F.R.K.: The Diameter of a Cycle Plus a Random Matching. Siam Journal on Discrete Mathematics 1, 328–333 (1988)
9. Bunke, H., Allermann, G.: Inexact graph matching for structural pattern recognition. Pattern Recognition Letters 1, 245–253 (1983)
10. Bunke, H., Shearer, K.: A graph distance metric based on the maximal common subgraph. 19, 255–259 (1998)
11. Dean, J., Ghemawat, S.: MapReduce: Simplied Data Processing on Large Clusters. In: Operating Systems Design and Implementation, pp. 137–150 (2004)
12. Deshpande, M., Kuramochi, M., Karypis, G.: Frequent Sub-Structure-Based Approaches for Classifying Chemical Compounds. In: IEEE International Conference on Data Mining, pp. 35–42 (2003)

13. Holder, L.B., Cook, D.J., Djoko, S.: Substucture Discovery in the SUBDUE System. In: Knowledge Discovery and Data Mining, pp. 169–180 (1994)
14. Inokuchi, A., Washio, T., Motoda, H.: An Apriori-Based Algorithm for Mining Frequent Substructures from Graph Data. In: Zighed, D.A., Komorowski, J., Żytkow, J.M. (eds.) PKDD 2000. LNCS (LNAI), vol. 1910, pp. 13–23. Springer, Heidelberg (2000)
15. Jin, C., Bhowmick, S.S., Xiao, X., Cheng, J., Choi, B.: GBLENDER: towards blending visual query formulation and query processing in graph databases. In: Proceedings of the 2010 ACM SIGMOD International Conference on Management of Data, SIGMOD 2010, pp. 111–122. ACM, New York (2010)
16. Jin, C., Bhowmick, S.S., Xiao, X., Choi, B., Zhou, S.: Gblender: visual subgraph query formulation meets query processing. In: SIGMOD Conference, pp. 1327–1330 (2011)
17. Kumar, R., Raghavant, P., Rajagopalan, S., Sivakumar, D., Tomkins, A., Upfal, E.: Stochastic models for the Web graph. In: IEEE Symposium on Foundations of Computer Science, pp. 57–65 (2000)
18. Malewicz, G., Austern, M.H., Bik, A.J., Dehnert, J.C., Horn, I., Leiser, N., Czajkowski, G.: Pregel: a system for large-scale graph processing. In: Proceedings of the 2010 ACM SIGMOD International Conference on Management of Data, SIGMOD 2010, pp. 135–146. ACM, New York (2010)
19. Melnik, S., Garcia-Molina, H., Rahm, E.: Similarity flooding: A versatile graph matching algorithm and its application to schema matching. In: ICDE, pp. 117–128 (2002)
20. Mongiovì, M., Natale, R.D., Giugno, R., Pulvirenti, A., Ferro, A., Sharan, R.: Sigma: a set-cover-based inexact graph matching algorithm. J. Bioinformatics and Computational Biology 8(2), 199–218 (2010)
21. Neumann, T., Weikum, G.: The RDF3X engine for scalable management of RDF data. The Vldb Journal 19, 19:91–19:113 (2010)
22. Padmanabhan, S., Chakravarthy, S.: HDB-Subdue: A Scalable Approach to Graph Mining. In: Pedersen, T.B., Mohania, M.K., Tjoa, A.M. (eds.) DaWaK 2009. LNCS, vol. 5691, pp. 325–338. Springer, Heidelberg (2009)
23. Pei, J., Han, J., Mortazavi-Asl, B., Pinto, H., Chen, Q., Dayal, U., Hsu, M.-C.: PrefixSpan,: mining sequential patterns efficiently by prefix-projected pattern growth. In: International Conference on Data Engineering, pp. 215–224 (2001)
24. Pelillo, M., Mestre, V.: Replicator Equations. Maximal Cliques, and Graph Isomorphism 11, 1933–1955 (1999)
25. Sun, Z., Wang, H., Wang, H., Shao, B., Li, J.: Efficient subgraph matching on billion node graphs. PVLDB 5(9), 788–799 (2012)
26. Suri, S., Vassilvitskii, S.: Counting triangles and the curse of the last reducer. In: World Wide Web Conference Series, pp. 607–614 (2011)
27. Tian, Y., Patel, J.M.: Tale: A tool for approximate large graph matching. In: ICDE, pp. 963–972 (2008)
28. Tong, H., Faloutsos, C., Gallagher, B., Eliassi-Rad, T.: Fast best-effort pattern matching in large attributed graphs. In: KDD, pp. 737–746 (2007)
29. Umeyama, S.: An Eigendecomposition Approach to Weighted Graph Matching Problems. IEEE Transactions on Pattern Analysis and Machine Intelligence 10, 695–703 (1988)
30. Valiant, L.G.: A bridging model for parallel computation, vol. 33, pp. 103–111. ACM, New York (August 1990)

31. Yan, X., Han, J.: gSpan: Graph-Based Substructure Pattern Mining. In: IEEE International Conference on Data Mining, pp. 721–724 (2002)
32. Yan, X., Yu, P.S., Han, J.: Graph indexing: A frequent structure-based approach. In: SIGMOD Conference, pp. 335–346 (2004)
33. Zager, L.A., Verghese, G.C.: Graph similarity scoring and matching. In: Applied Mathematics Letters, vol. 21, pp. 86–94 (2008)

Complex Network Characteristics and Team Performance in the Game of Cricket

Rudra M. Tripathy[1], Amitabha Bagchi[2], and Mona Jain[2]

[1] Silicon Institute of Technology, Bhubaneshwar, India
[2] Indian Institute of Technology, Delhi, India

Abstract. In this paper a complex network model is used to analyze the game of cricket. The nodes of this network are individual players and edges are placed between players who have scored runs in partnership. Results of these complex network models based on partnership are compared with performance of teams. Our study examines Test cricket, One Day Internationals (ODIs) and T20 cricket matches of the Indian Premier League (IPL). We find that complex network properties: average degree, average strength and average clustering coefficient are directly related to the performance (win over loss ratio) of the teams, i.e., teams having higher connectivity and well-interconnected groups perform better in Test matches but not in ODIs and IPL. For our purpose, the basic difference between different forms of the game is duration of the game: Test cricket is played for 5-days, One day cricket is played only for a single day and T20 is played only for 20 overs in an inning. In this regard, we make a clear distinction in social network properties between the Test, One day, and T20 cricket networks by finding relationships between average weight with their end point's degrees. We know that performance of teams varies with time - for example West Indies, who had established themselves as the best team during 1970s now is one of the worst teams in terms of results. So we have looked at evolution of team's performances with respect to their network properties for every decade. We have observed that, the average degree and average clustering coefficient follow similar trends as the performance of the team in Test cricket but not in One day cricket and T20. So partnership actually plays a more significant role in team performance in Test cricket as compared to One day cricket and T20 cricket.

Keywords: Complex Networks, Social Networks, Cricket, Partnership.

1 Introduction

Networks or graphs provide a representation of highly complex phenomena in a simpler form allowing us to understand the system in a better way. Obviously, the applicability of this approach depends on the ability to identify meaningful interacting units and the relationships connecting them. These complex networks are analyzed in terms of some static properties based on degree and connectivity. Cricket is probably the only sport where a dyadic relationship exists among a

V. Bhatnagar and S. Srinivasa (Eds.): BDA 2013, LNCS 8302, pp. 133–150, 2013.

group of players in the form of batting partnership (runs scored between two players together). Out of 11 players of a batting team only two are batting at a given time. So it is quite natural to think of cricket as a dyadic social network. It is common wisdom that batting partnerships, sometimes called "offensive relationships", play a significant role in performance (win over loss ratio) of the team in cricket. In order to put this wisdom on a rigorous basis we modeled cricket as a social network based on partnership and studied the network properties of the cricket network. Our main observation is that team performance in Test cricket definitely depends on the presence of players who form many strong partnerships, but in the shorter forms of the game this effect is weakened. While this result is what we might expect, our contribution is methodological in that we describe a set of metrics by which this may be established.

Experiments and overview of results: Here a weighted network model has been considered to represent each team where each cricket player is a node in the network. Two nodes are connected by an edge if they have scored runs in partnership and the weight of the edge is given by the ratio between total number of runs scored and number of partnerships. Different social network metrics are then applied on the created networks to get insight into the nature of social relationships between players of a team and the effect of these relationships over the performance of a team. Some of others interesting observations from the cricket network are as follows:

- Degree distribution as well as strength distribution follows power law [1,2,3,4,5] and it also demonstrates the *small world* phenomena [6,7,3]
- There are few hubs in the partnership graphs. They represent popular (higher degree) batsmen of their respective countries like Sachin Tendulkar and Ricky Ponting.
- In the cricket network, clustering coefficient of nodes decreases with their degrees. This behavior is explained as the larger-degree players partner with players who were on the team in different time periods.
- The cricket network satisfies disassortative properties. So in the cricket network, higher degree nodes have majority of neighbors with lower degrees, whereas opposite holds for low-degree vertices
- We found that clustering coefficient and average degree are directly related to the performance of the team, i.e., a team having higher connectivity and well-interconnected groups performs better.
- We show that partnership plays a more significant role in Test cricket than One day and T20 for team performance.

Dataset Description: The partnership data were collected from the Stats-Guru utility on the ESPNCricinfo website [8]. Advance filter tab was used to obtain the overall partnership summary of all the Test matches played between 1st Jan. 1950 to 31st Dec. 2008 and One day matches played between 1st Jan. 1971 to 15th July 2009 by each of the desired team. We have considered only those partnerships between two players who have scored more than 30 runs in three such matches in Test cricket and scored more than 20 runs in five such matches in One day cricket. The reasons for taking different measures for Test and One day cricket are - One day cricket is played only for a single day, whereas Test cricket is played for five days and the number of One day matches played in a calendar year is more than that of Test matches. For the shortest format of the cricket, i.e., T20, we have used Indian Premier League (IPL) data. The Cricinfo website is used to gather the partnership data for all six seasons in which IPL tournament has been played, i.e., from 2008 to 2013. Each of the team in IPL has nearly played 100 matches. Therefore, we have not kept any restriction on the minimum partnership score or minimum number of matches for the IPL data. Some part of our analysis was done in 2010, particularly for the One day and Test cricket. Hence for Test and One day cricket we have used the old crawled dataset.

The remainder of the paper is structured as follows: Section 2 contains related works, Section 3 deals with background and motivation for this work, Section 4 we discuss how to model cricket as a complex network, Section 5 we relate the properties of cricket network with respect to the performance of the corresponding team and finally in Section 6 we draw some conclusions and highlight future work.

2   Related Work

Large complex systems, such as biological, social, technological and communication systems, have been analyzed using network models. The Internet [5], the World Wide Web [1], the online social networking services [2], the scientific collaboration networks (SCN) [4] are just to name a few. A number of studies have been done to investigate the topological characteristics of many networks. Milgram's work [7] giving concepts like *six degrees of separation* and *small world* are the most popular in this area. Measures such as *degree* of a node, *average nearest neighborhood degree* and *clustering coefficient* have been well studied for the unweighted network [9]. Few studies have also been done in weighted networks. In [10], Newman proposed that properties of weighted networks could be studied by mapping them to unweighted multigraphs. In [11], Barrat et al. have given a quantitative and general approach to understand the complex architecture of real weighted networks. The analysis of team performance of sports leagues using complex network properties can be found in the literature. Vaz De Melo et. al. [12] have analyzed the National Basketball Association (NBA) database using complex network metrics, such as clustering coefficient and node degree. They have shown that the NBA network can be characterized as a small-world network and the degree distribution of players follows a power law. They have also proposed models for predicting the behavior of a team based on box-score statistics and complex network metrics. Results showed that only the box-score statistics cannot predict the correct behavior. But a mixed model (the network properties together with the box score statistics) can give more accurate results. In another study, Onody and De Castro [13] studied the Brazilian National Soccer Championship statistics from 1971 to 2002 by forming a bipartite network between the players and the clubs and they found that the connectivity of players had increased over the years while the clustering coefficient declined. This means, the players' professional life is growing longer and the players' transfer rate between clubs is going up. One of the major studies in the game of cricket is done by Uday Damodaran. He analyzed the ODI batting performance of Indian players from 1989 to 2005, using stochastic dominance rules [14]. His study includes 4 specialist batsmen and 1 bowler and the results found were intuitive, such as Sachin Tendulkar proved to out-stand other players while Rahul Dravid dominated others till 20 overs. In their work [15], Allsopp and Clarke have rated cricket teams using multiple linear regression for both One day and Test. They have found Australia and South Africa are more rated teams than others. In this work, we have modeled cricket as a complex network and analyzed cricket using complex network properties. In our opinion similar work is not present in the literature. This is the first time that the game of cricket has been put into a social network framework.

3   Introduction to Cricket

The game of cricket originated in England. The game is played between two teams of eleven players each. The teams comprises some batsmen, some bowlers and a wicket-keeper. There are three forms of the game in cricket: Test matches, ODIs and T20. Test match cricket is the longest form of match, and is played over five days, with three sessions of two hours on each day. The first Test match was played between England and Australia in 1877 and since then a total of 1920 matches have been played up to 20th July 2009. A One day or limited-over match lasts for only a single day, where each team bats only once and their innings are limited to a fixed number of overs, usually fifty. The first One day international was played in Melbourne, Australia, in 1971, since then total number of 2861 number of matches have been played. One day cricket is more popular than Test cricket because of aggressive, risky and entertaining batting, which often results in a close finish. T20 is the shortest type of cricket as only 20 overs are played per inning. Indian Premier League(IPL) is a league for T20 championship in India. Both Indian and International players take part in the tournament. The first season of IPL was played in 2008 involving a total of 8 teams. Since then six IPL tournaments have been played.

Partnership In cricket two batsmen always bat together which is called partnership. Only one batsman is on strike at any time. If one of them got out then the partnership is broken. The measures which are used to describe a partnership are the number of runs scored, the duration of the partnership and the number of balls faced. Partnership plays a key role in match outcome in Test cricket as well as in One day cricket.

3.1   Motivation

As we know that the performance of a team for a particular period is measured by the number of matches they win in that period. But perhaps the better

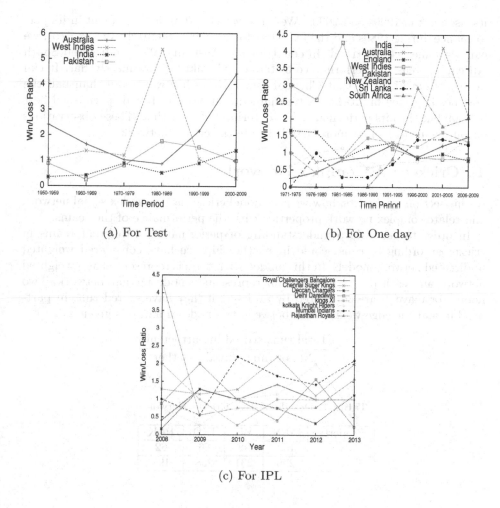

(a) For Test

(b) For One day

(c) For IPL

Fig. 1. Performance of teams

measure is the ratio between the number of matches won and the number of matches lost. So we did a performance study of some important cricket-playing countries (those that played more than 400 matches) in both form of cricket based on their win/loss ratio.

The performance of some selected Test playing countries from 1951 to 2009, One day playing counties from 1971 to 2009, and IPL playing teams from 2008 are shown in Fig. 1(a), Fig. 1(b) and Fig. 1(c) respectively. As clear from Fig. 1(a), Australia and West Indies are two countries whose win/loss ratio has crossed 4.5, while other countries have stayed below 2.0. For One day cricket, similar things can be observed in Fig. 1(b), Australia and West Indies have higher win/loss ratios than others. Some other interesting things can be seen in Fig. 1(a) like the dip in Australia's performance from 1950s to 1980s and then its rise back in 1990s and 2000s, West Indies showed an improvement in its performance during 1980s and the fell back during 1990s. Similarly, in Fig. 1(b), we can see some definite trends in countries like Australia, West Indies and South Africa, whereas the performance of India and Pakistan are mostly flat. It can be seen from Fig. 1(c) that Chennai Super Kings has won more than half the matches it played in all the six tournaments. Rajasthan Royals performed best in 2009 but then its performance has deteriorated after that. These observations encourage us to look at more details inside the cricket network.

4   Cricket : A Complex Network

In this section we discuss, how we can model cricket as a complex social network and relate complex network properties with the performance of the teams.

In order to get a deep understanding of performance of different teams in cricket according to runs scored in partnership, we have considered weighted undirected network models. In this model, each team is represented as a weighted network and each player in the team represents a node in that network. Two nodes (or players) are connected by an edge, if they have scored runs in partnership and the edge weight $w_{u,v}$ between two nodes u and v is given as

$$w_{u,v} = \frac{\text{Total runs scored in partnership}}{\text{No. of Inns played together}}$$

Table 1. Details of Test Cricket Network

Property	AUS	WI	IND	PAK
v	102	98	86	62
e	296	231	238	193

Table 2. Details of One day Cricket Network

Property	AUS	ENG	IND	NZ	PAK	SA	SL	WI
v	55	40	40	44	48	27	35	36
e	122	88	115	115	134	81	94	116

Table 3. Details of IPL Cricket Network

Property	RCB	CSK	DC	DD	KingsXI	KKR	MI	RR
v	65	41	66	58	57	56	51	58
e	226	142	253	235	236	207	224	251

The main objective in this work is to view cricket from a social network perspective based on partnership, because partnership can be thought of as a social relationship, i.e., more the number of runs scored between two players in partnership, closer their social relationship. An undirected, weighted graph is created for each country. The details about the complexity of the Test network is given in Table-1, One day network in Table-2, and IPL in Table-3.

We have done all our experiments using Networkx [16], which is a Python-based package for creation, manipulation, and study of the structure, dynamics, and function of complex networks. Complex network structures have two basic set of properties, topological and structural. To characterize cricket as a complex network, we looked at the two set of properties for the cricket network.

Topological Characterization: Here we have considered degree and strength distribution of nodes. The resultant graph for the degree distribution for Test cricket network is shown in Fig. 2(a). Here k is referred as degree and $P(k)$ denotes the degree distribution, i.e., probability that a given node is having degree k.

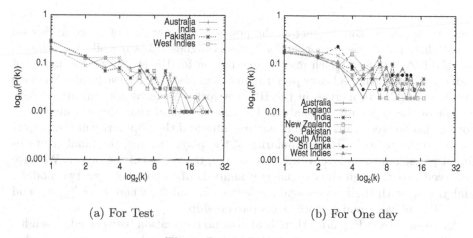

(a) For Test   (b) For One day

Fig. 2. Degree Distribution

Here a degree of a node is determined by number of partners in it, i.e., number of team members with whom the node player has scored more than 30 runs in at least 3 innings. This distribution can be approximated by a power law behavior and the power law exponents (as described by Clauset et al. in [17]) for Australia, India, Pakistan and West Indies are found to be 3.50, 2.79, 2.89 and 2.37 respectively. Hence we can see a big difference in power law exponent between Australia and West Indies which reflects in their performance. Similar results are obtained for One day cricket (Fig. 2(b)). But, here degree of a node determines - number of partners with whom the player has scored more than 20 runs in at least 5 innings. This distribution also follows power law behavior. Australia has more players having smaller degrees as compared to other countries, which means a lot of Australian players are capable of creating partnership. So we can say that, Australia is not a *few men team*, which makes it a stronger

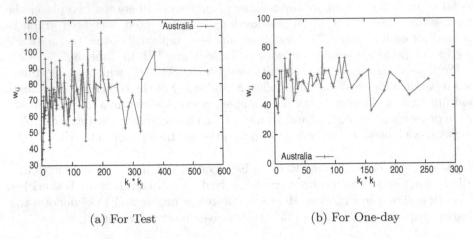

(a) For Test   (b) For One-day

Fig. 3. Average weight as a function of end-point degrees for AUSTRALIA

team than others. But looking at the power-law exponents of all countries we found that, power law exponent for India is highest (3.50), while West Indies has the lowest (1.80), which may be one reason for the steady performances of India throughout and a heavy performance degradation of West Indies. We have also seen similar properties in the IPL network. We have not considered only weight of the edges in our analysis, because we found that there is almost no correlation between runs scored in partnerships and the popularity of the players for Australia network. Here popularity of the player means, the number of different partnerships a player is involved in, i.e., the degree of the player. We have observed this through the dependency analysis of weight between two nodes i and j, $w_{i,j}$, with their corresponding degrees k_i and k_j, where $k_i = \sum_k a_{ik}$ and $a_{ik} = 1$, if player i and k scored runs partnership.

As we can see in Fig. 3(a), there is almost no correlation between edge weights and their end-point degrees except for higher $k_i * k_j$ values, are having higher $w_{i,j}$ values. But in ODIs Fig. 3(b), most of the edge weights are between 50 to 70 regardless the values of $k_i * k_j$.For IPL also, most of the edge weights are between 15-30 irrespective for the end-point degrees. From these observations, we can conclude that, two higher degree players involve in higher partnership in Test cricket than One day cricket and T20, i.e., the role of popular (higher degree) players like RT Ponting, SR Waugh, AR Border are more significant in Test than One day and T20. In One day and T20 cricket almost all players have equal role to play in team performance.

Hence without considering the individual weight we consider group weights, which is the sum of all adjacent edges weight. So we looked at the extended definition of vertex degree k_i in terms of vertex *strength* s_i, which is defined as

$$s_i = \sum_{j=1}^{N} a_{ij} w_{ij} \qquad (1)$$

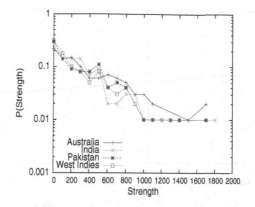

Fig. 4. Strength Distribution for Test Cricket

Fig. 4 shows the probability distribution $P(s)$ that a vertex has strength s in the Test cricket. The functional behavior of this distribution is very similar to that of degree distribution $P(k)$ in Test and One day cricket. But while looking at the power law coefficient, we found that India has the minimum (1.63) power law coefficient. That may be reason due to which most of the time between 1950 to 2009, India's performance being least among all the four countries.

Structural Organization: In the cricket network, there may exist well connected groups of players - since two players who play for early wickets will be more connected to each other than those playing for first and the last wickets. In order to find the local group cohesiveness and correlations between the degree of connected vertices in the cricket network, we analyzed two social network characteristics clustering coefficient and assortativity.

For any node v_i, the clustering coefficient c_i is defined as the fraction of connected neighbors of v_i [6]. In general, this metric tells us how well a typical node's neighbors are connected in the network. The clustering coefficient of the network is the average clustering coefficient $C = N^{-1} \sum_i c_i$. Since the graphs involved here are weighted graph, we consider here the *weighted clustering coefficient*. In [11] this was defined as

$$C_i = \frac{1}{s_i(k_i - 1)} \sum_{j,h} \frac{(w_{ij} + w_{ih})}{2} a_{ij} a_{ih} a_{jh} \qquad (2)$$

We have defined C^k as average weighted clustering coefficient of all vertices having degree k. In all three different forms of cricket (Test, One day, and IPL), the networks show a decaying C^k. That means players with fewer partners usually make a well defined group in which all the players know each other. On the other hand, players with large degrees, know players from different groups, who in turn does not know each other. This is largely due to the partial change in the list of team players at certain times when a few old players are replaced with

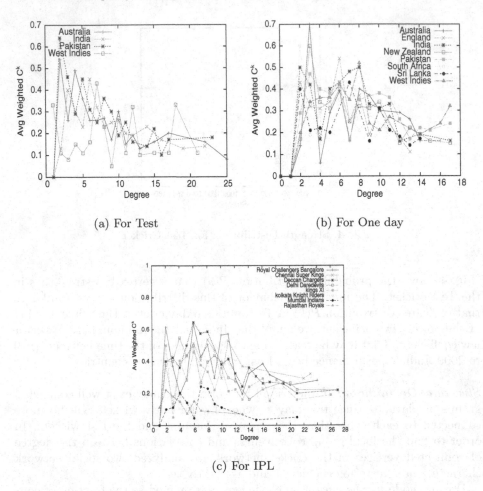

(a) For Test

(b) For One day

(c) For IPL

Fig. 5. Average weighted clustering coefficient as a function of k

new players. The value average weighted C^k is lower for Rajasthan Royals and Mumbai Indians as they have the larger number of partnerships with less number of players which can also be verified from Table-3 and is highest for Royal Challengers Bangalore.

To know the interconnection between nodes, we have looked at the average degree of nearest neighbors, $k_{nn}(k)$. This quantity is called as the joint degree distribution (JDD), which gives insight into the structure of the neighborhood of a node with degree k. An increasing $k_{nn}(k)$ indicates a tendency of high degree nodes tend to connect to other high degree nodes. This property is popularly known as the *assortative mixing* [18,19]. A decreasing nature of this quantity defines *disassortative mixing*, where high-degree vertices have majority of low-degree neighbors. Since the networks used are weighed networks, we consider the *weighted average nearest-neighbor degree* [11,18,19]. This is defined as

$$k_{nn,i}^{w}(k) = \frac{1}{s_i} \sum_{j=1}^{N} a_{ij} w_{ij} k_j \qquad (3)$$

If $k_{nn,i}^{w}(k) > k_{nn,i}(k)$ then edges with higher weights point to neighbors with larger degree and vice-versa. In both cases of Fig. 6(a) and (b), $k_{nn,i}^{w}(k)$ is plotted

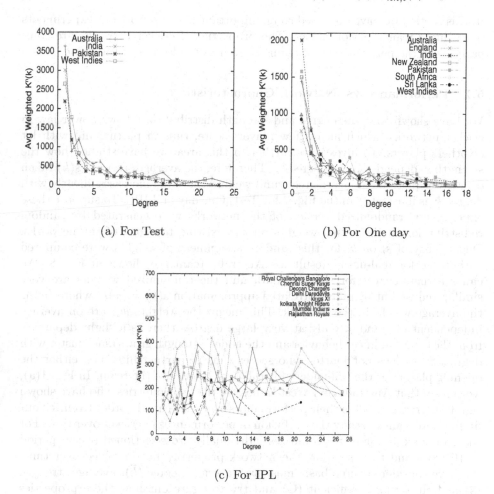

(a) For Test

(b) For One day

(c) For IPL

Fig. 6. Average weighted nearest-neighbor degree as a function of k

as a function of k, which decays exponentially. This means the cricket network for the case of Test and One day cricket exhibit a *disassortative mixing model* in which players with higher degrees have majority of low-degree neighbors. That is the probability that two popular (higher degree) players score runs in partnership is less than that of one popular and one non-popular (lower degree) player. However, for the case of IPL (Fig. 6(c)), this coefficient does not decrease exponentially and the reason for this being that IPL inning is played only for twenty overs and thus it gives less chance to the lower degree players to do batting. From the above discussion now it is clear that, indeed cricket networks follow standard complex network characteristics.

5   Observation from Cricket Network

In this section we have discussed some important findings from our experiments to evaluate teams' performances. The properties of cricket networks are compared with the performances of teams as well as the format of the game.

5.1   Performance vs. Network Characteristic

We have shown that the degree and strength distribution follows power law in cricket networks, which means few players score, runs in partnership with lot of others players. To investigate further in this area, we have studied how the strength s_i depends on the degree k_i. Therefore, the average strength $s(k)$ taken over k was plotted against degree k and we find that $s(k)$ increases linearly with degree k as illustrated in the Fig. 7 for Test, One day and IPL. To support these experiments, randomized version of the networks were generated by random redistribution of the actual weights on the existing topology of the networks. Dependency of s_i on k_i for this random assignment of weights were compared with those for real-data. Result for Australia team are shown in Fig. 8. We can see that, curves for the real data and the randomized weights are very similar and well fit by an uncorrelated approximation $s(k) = \langle w \rangle k$, where $\langle w \rangle$ is the average weight in the network. This means the weights w_{ij} are on average independent of i and j. Only at very large degrees there is a slight departure from the expected linear behavior and the nodes' strength grows much faster with degree. So it does not matter who scores runs in partnerships, i.e., either the opening players or the tail-enders for the performance of the team. In Fig. 1(a), we can see that Australia and West Indies are the two countries who have shown significant rise and fall in their performances over the years. In order to enrich our findings, we want to study the evolution of performances in cricket over time. For that, we look at these two countries individually cross-sectioned over a period of 10 years and then see how the network properties have evolved over time. We have considered three basic measures: average degree ($\overline{k}$), average strength ($\overline{s}$) and clustering co-efficient (C) and try to figure out how these properties behave with time evolution. The mean connectivity $\overline{k}$ or average degree for the Australia and West Indies are shown in Fig. 9(a). The findings are very intuitive and interesting. We have observed that the average degree of the nodes follow similar behavior that of the performances of the corresponding countries shown in Fig. 1(a). Next we investigate if the average strength, $\overline{s}$, of nodes follow the same suit as the average degree or not. Fig. 9(b) shows the average strength of both the countries in different time periods. As expected, we find a similar trend here as well. The behavior here is closer to the performance trends as compared to average degree. Lastly we try to find out the trends of evolution for average

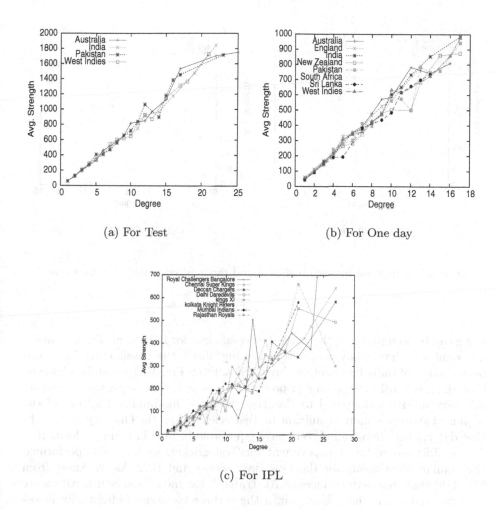

(a) For Test

(b) For One day

(c) For IPL

Fig. 7. Strength vs. Degree

clustering coefficient. Same behavior is shown in this case also. As clear from Fig. 9(c), we can see the striking similarities between team's performance and its average clustering coefficient over time.

5.2   Formats of the Game vs. Network Characteristics

Here we discuss some interesting findings related to the difference between Test, ODI and T20 cricket using complex network properties. In Fig. 4, we can clearly see that, Australia has more players having smaller degrees compared to other countries in One day cricket, which is not seen in Test Cricket. It follows that; a lot of players are capable of creating partnership for Australia in One day cricket. This probably makes Australia a stronger team in ODIs. Another interesting

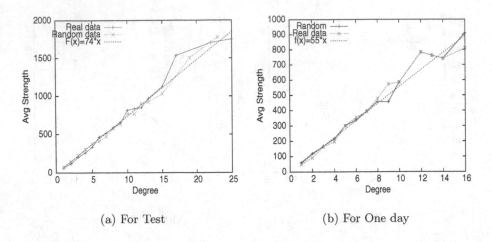

(a) For Test   (b) For One day

Fig. 8. Comparison between real data and randomized weights for Australia

thing can be seen in Fig. 4, the power law coefficient for India is maximum among all countries, that makes a strong reasoning about the consistently improved performance of India team in One day cricket. From Fig. 4(b), it can be observed that the relationship between end points degrees and their respective weight do not vary much as compared to the Test cricket. This implies the role of the popular batsmen is more significant in Test cricket than the One day cricket. In One day cricket almost all players have equal role to play in team performance. As we did time evolution experiment for Test cricket, we have also performed the similar experiments for the One day cricket and IPL. As we know from Fig. 1(b) that, the performances of Australia, West Indies and South Africa are very fluctuating, we have looked into these three countries individually cross-sectioned over a period of 5 years and then shown how the network properties have evolved over time. The results are given in Fig. 10. But here the network properties like degree, strength and clustering coefficient do not follow the same trend as the performance of the team. But the number of nodes and edges increase as the performance of the team increases. From this we can conclude that the partnership plays more crucial role in Test cricket than in One day cricket.

In the case of IPL the average strength, Fig. 11(b), follows the pattern similar to the performance of the team. The evolution of average degree and clustering coefficient with time does not follow the same trend as the performance of the team. The reason for this can be explained as, some of the players played for different franchises in different seasons and there are lot of changes made in the batting order. The change in batting order matters a lot in case of IPL as it being a T20, only 20 overs are played and hence not all players get a chance to do batting.

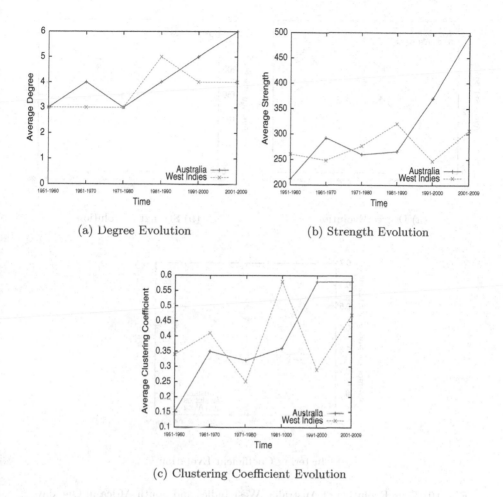

(a) Degree Evolution

(b) Strength Evolution

(c) Clustering Coefficient Evolution

Fig. 9. Time Evolution of Australia and West Indies in Test

6   Conclusion

In this work, we have modeled cricket as a complex social network, by showing that degree distributions and strength distributions follow power-law and also decrease in clustering coefficient with degree and having disassortative properties. We also found few hubs in the networks; interestingly these hubs in the networks are the popular batsmen of their respective countries like Sachin Tendulkar and Ricky Ponting. We have also analyzed the cricket network with respect to the performance of teams in all the three formats of cricket, Test, ODIs and T20. We found that, the edge weights are largely independent of end-points degree, i.e., there is almost no correlation between the partnership runs scored and the pair of partners, except the case for very low-degree players who in general score

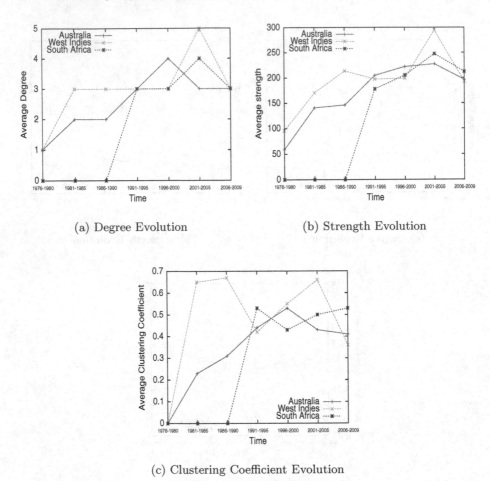

(a) Degree Evolution   (b) Strength Evolution

(c) Clustering Coefficient Evolution

Fig. 10. Time Evolution of Australia, West Indies and South Africa in One day

less and very large-degree players score more in Test cricket. Whereas in ODIs and T20, edge weights mostly have a constant relationship with end-point degrees. Hence in Test cricket, larger degree players like Sachin Tendulkar, Rahul Dravid are more significant than others but in One day cricket and IPL almost all players are equally important to the team. We have shown that, the clustering coefficient of nodes decrease in the cricket networks. Hence hubs in the networks have very dispersed neighborhoods while the low-degree vertices form well-connected groups. This behavior is explained as the larger-degree players partner with players who were on the team in different time periods. We have shown that, the average degrees of nearest neighbors of nodes with degree k decays exponentially with k for Test and ODIs, which signifies a disassortative behavior of the network, but generally real social networks satisfy assortative behavior. So in cricket network, high degree nodes have majority of neighbors with low degrees, whereas opposite holds for low-degree vertices. Interestingly

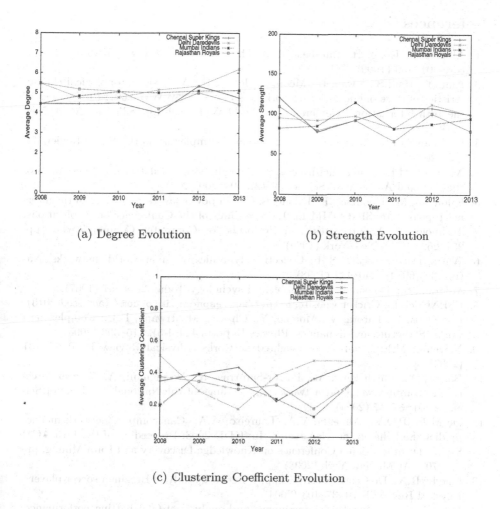

(a) Degree Evolution   (b) Strength Evolution

(c) Clustering Coefficient Evolution

Fig. 11. Time Evolution in IPL

we found similar trend follows in all forms of cricket. Due to large difference in the performance of Australia and West Indies in Test matches, we have observed the performance of teams with time evolution. We get very surprising results as we see that the average degree and the clustering coefficient follow the same trend as the performance of the country for Test cricket. But while doing the similar experiment for One day and T20 cricket, we have got completely opposite results, i.e., these properties are no way related to the performance of the team. Hence, the most important conclusion is, partnership plays more significant role in Test matches than in One day and T20 matches. As future works, one can go further in the time evolution of network characteristics. As these properties are directly correlated to the teams performance, one can think for developing models in this frame-work to predict the match outcomes.

References

1. Albert, R., Jeong, H., Barabasi, A.L.: The diameter of the world wide web. Nature 401, 130 (1999)
2. Ahn, Y., Han, S., Kwak, H., Moon, S., Jeong, H.: Analysis of topological characteristics of huge online social networking services. In: WWW 2007: Proceedings of the 16th International Conference on World Wide Web, pp. 835–844. ACM, New York (2007)
3. Newman, M.E.: The structure and function of complex networks. SIAM Review 45, 167 (2003)
4. Newman, M.E.: The structure of scientific collaboration networks. Proceedings of the National Academy of Sciences 98(2), 404–409 (2001)
5. Faloutsos, M., Faloutsos, P., Faloutsos, C.: On power-law relationships of the internet topology. In: SIGCOMM 1999: Proceedings of the Conference on Applications, Technologies, Architectures, and Protocols for Computer Communication, pp. 251–262. ACM, New York (1999)
6. Watts, D.J., Strogatz, S.H.: Collective dynamics of 'small-world' networks. Nature 393(6684), 440–442 (1998)
7. Milgram, S.: The small world problem. Psychology Today 2, 60–67 (1967)
8. ESPNCricinfo: Cricket website, http://www.espncricinfo.com/ (accessed 2013)
9. Boccaletti, S., Latora, V., Moreno, Y., Chavez, M., Hwang, D.U.: Complex networks: Structure and dynamics. Physics Reports 424(4-5), 175–308 (2006)
10. Newman, M.E.: Analysis of weighted networks. Physical Review E 70, 56–131 (2004)
11. Barrat, A., Barthelemy, M., Pastor-Satorras, R., Vespignani, A.: The architecture of complex weighted networks. Proceedings of the National Academy of Sciences 101, 37–47 (2004)
12. De Melo, P.O.V., Almeida, V.A., Loureiro, A.A.: Can complex network metrics predict the behavior of NBA teams? In: KDD 2008: Proceeding of the 14th ACM SIGKDD International Conference on Knowledge Discovery and Data Mining, pp. 695–703. ACM, New York (2008)
13. Onody, R.N., De Castro, P.A.: Complex network study of Brazilian soccer players. Physical Review E 70, 37–103 (2004)
14. Damodaran, U.: Stochastic dominance and analysis of ODI batting performance: The Indian cricket team, 1989-2005. Journal of Sports Science and Medicine 5(4), 503–508 (2006)
15. Allsopp, P.E., Clarke, S.R.: Rating teams and analysing outcomes in One-day and Test cricket. Journal of The Royal Statistical Society Series A 167(4), 657–667 (2004)
16. NetworkX: Networkx documentation, http://www.networkx.lanl.gov/ (accessed 2013)
17. Clauset, A., Shalizi, C.R., Newman, M.E.: Power-law distributions in empirical data. SIAM Review 51(4) (November 2009)
18. Leung, C.C., Chau, H.F.: Weighted assortative and disassortative networks model (2006)
19. Chang, H., Su, B., Zhou, Y., He, D.: Assortativity and act degree distribution of some collaboration networks. Physica A: Statistical Mechanics and its Applications 383(2), 687–702 (2007)

Visualization of Small World Networks Using Similarity Matrices

Saima Parveen and Jaya Sreevalsan-Nair

International Institute of Information Technology Bangalore,
26/C Electronics City, Bangalore 560100, India
saima.parveen@iiitb.org, jnair@iiitb.ac.in
http://www.iiitb.ac.in

Abstract. Visualization of small world networks is challenging owing
to the large size of the data and its property of being "locally dense but
globally sparse." Generally networks are represented using graph layouts
and images of adjacency matrices, which have shortcomings of occlusion
and spatial complexity in its direct form. These shortcomings are usually
alleviated using pixel displays, hierarchical representations in the graph
layout, and sampling and aggregation in the matrix representation. We
propose techniques to enable effective and efficient visualization of small
world networks in the similarity space, as opposed to attribute space, using similarity matrix representation. Using the VAT (Visual Assessment
of cluster Tendency) algorithm as a seriation algorithm is pivotal to our
techniques. We propose the following novel ideas to enable efficient hierarchical graphical representation of large networks: (a) parallelizing VAT
on the GPUs, (b) performing multilevel clustering on the matrix form,
and (c) visualizing a series of similarity matrices, from the same data set,
using parallel sets-like representation. We have shown the effectiveness
of proposed techniques using performance measurements of parallel implementation of VAT, results of multilevel clustering, and analyses made
in case studies.

1 Introduction

Consider a time series of coauthorship network. Graph layout of such a network
can show the "hubs," or nodes with high centrality, e.g., which may indicate fac-
ulty members heading labs or research groups. Additionally, it will be insightful
to find significant events in the network, e.g., when two authors who worked together earlier have stopped publishing together or two authors who did not work
together are collaborating currently. However it will be difficult to infer temporal
changes in the network in a single view. Visualizing large networks tend to be a
challenge which can be addressed by constructing meaningful hierarchical structures. We propose using similarity-based clustering in the network to identify
features such as temporal or spatial events and to construct hierarchical levels
of detail.

Similar to [21], we will be using the terms "graph", "vertices", and "edges" to
refer to the topological structures with no associated attributes, and "network",

V. Bhatnagar and S. Srinivasa (Eds.): BDA 2013, LNCS 8302, pp. 151–170, 2013.
© Springer International Publishing Switzerland 2013

"nodes", and "links" to refer to structures related to a graph with attributes, respectively. Elements of an adjacency matrix correspond to vertices, and those of a similarity matrix correspond to nodes.

Our work is based on the premise that similarity functions can be considered to be transformations on the adjacency matrix and can be further used for clustering to manage large data. Several analyses performed on a network, e.g. using similarity functions, centrality measures, etc. can be reduced to being a transformation on the adjacency matrix [27], [6], and hence it is implicitly a transformation on the graph itself. These transformations can be linear or nonlinear. This serves as a motivation for our work, as graphical representation of these transformations can help us gain more insights into the data and our work can extend beyond our current scope of analysis using similarity functions, e.g., analysis based on centrality measures can be made from matrix representations.

For this study, we have focused on small world networks in its undirected form characterized by a local structure where two connected nodes can have several neighbors. Extensive studies on small world networks can be found in [47], [4]. Applying a similarity function, which is symmetric, on an adjacency matrix yields a normalized symmetric matrix called similarity matrix. Though our work can be extended to any network and any normalized symmetric matrix which is characteristic of the network, the performance of our techniques will vary depending on the properties of the network and the transformation applied on the adjacency matrix to obtain the concerned matrix.

We have found that in the information visualization community, matrix visualization has been routinely applied for adjacency matrices for undirected graphs. Adjacency matrices are generally visualized as a binary shaded representation of a matrix, indicating presence or absence of edges between vertices using white and black colors, respectively. Visualizing grayscale matrix representations for similarity matrices has been done before, such as the shaded similarity matrices and variants of the Visual Assessment of cluster Tendency (VAT) algorithm [3], where permuted or seriated similarity matrix representations are used for finding clusters. Using VAT for seriation of the similarity matrix is pivotal to our work. VAT is considered equivalent to a single linkage hierarchical algorithm, which is significant as we use the seriated similarity matrix to construct multiple levels of detail in a hierarchical fashion, for which an underlying single linkage algorithm is apt. Since VAT has a quadratic time complexity, it is inefficient for large scale data sets. Hence we propose a parallel implementation of VAT on the GPUs using CUDA [41].

Multilevel clustering has traditionally been done on the node-link diagrams using nearest neighbor consideration, which is satisfied by VAT owing to its relation to single linkage algorithm. Multilevel clustering performed on the network has been carried out in ways which are specific to research communities: (a) in information visualization community, multilevel clustering of the graph is used for obtaining multiple levels of detail and for further processing pertaining to building focus+context representations, and (b) in machine learning community, one level of clustering is done using similarity functions. We have not found any work pertaining to multilevel clustering in a network combining the two approaches, i.e., obtaining multiple levels of detail using similarity functions. Finding meaningful functions to persist clustering in multiple levels of detail is in itself a challenging problem and is beyond the scope of our current work.

A series of similarity matrices can be generated from a data set in several ways, e.g., every time stamp in time series data, application of different similarity functions on the same data, or different subspace clustering in multivariate data. Finding two or more authors in the same cluster in a coauthorship network over a period of time may indicate that they are working very closely to each other owing to their employment at the same workplace, or having the right set of complementary expertise in pursuing research on the same topic or project. These authors not coauthoring a paper together after a while might indicate that they no longer work together owing to change of employment or change in research interests, etc. Information about such an event becomes very apparent from the clustering tendencies found in similarity matrix series. Tracking membership of objects in clusters across a series of similarity matrices can help in finding trends and patterns, and also in understanding the underlying varying properties of the data. The idea is that if we observe that two objects are present in the same clusters in multiple instances in the series of similarity matrices, we can conclude that the objects are positively correlated with respect to the underlying properties of the different instances. While there is existing research on visualizing similarity matrices for small world networks, we have not found any visual analysis for a series of similarity matrices. This is unique to our work, where we have adapted parallel sets representation to see clustering trends in a series of similarity matrices.

Summarizing, our contributions are as follows:

1. We propose a GPU-based parallel implementation of VAT (pVAT), which is an optimal parallel implementation of VAT using CUDA based on Borůvka's algorithm [44].
2. For data simplification we use the VAT images to perform multilevel clustering, i.e., create multiple levels of detail by recursively clustering the nodes and merging the nodes in the clusters to form new nodes.
3. We propose parallel sets-like representation for tracking the cluster membership of objects in series of similarity matrices generated from the same data, i.e., to view how constituency of the clusters changes across the series, and apply this technique on visualizing small world networks.

2   Related Work

Our design choices have been the result of choosing (a) matrix visualization over graph layouts for small world networks, and (b) VAT algorithm for seriation of similarity matrices. Network visualization and analysis, and that pertaining to small world networks are very active areas of research. Though it will be impossible to refer to the huge body of related literature in this section, we have listed the relevant research ideas that have influenced our work. Though one level of clustering in similarity matrix is routinely used in data mining applications, we did not find any related work on multilevel clustering performed on a similarity matrix. Similarly we have not found any work on analysis of a series of similarity matrices of a multivariate data set, including network data.

Small World Network Visualization:
Watts and Strogatz [47] have explained the characteristics of small world networks based on the small world phenomenon, also known as six degrees of separation [13]. There are several multiscale, multilevel visualizations of graph layout of large-scale small world networks [2], [1], [15], [9], which are scalable applications. These works largely exploit the property of small world networks, i.e. globally sparse and locally dense layout [4], using graph layouts.

Ghoniem et al. [11], [12] have analyzed comparison of graph visualization techniques between node-link diagrams and matrix based representations using controlled experiments. Though the experiments conducted in [11] primarily were for low-level readability tasks and not specifically for social networks, the results are applicable to small world networks. Using experimental results in [11], Ghoniem et al. have shown that node-link diagrams are suited for smaller graphs and graphs that were "almost trees", but performed very poorly for almost complete graphs and denser networks. The idea is that, given the property of small world networks of being globally sparse and locally dense and that matrix visualization is less sensitive to density variations, the matrix visualization will be better suited for small world networks for clarity.

Matrix Visualization of Networks:
Henry and Fekete [19] have proposed a network visualization system, MatrixExplorer, which uses two representations, namely, node-link diagrams and images of adjacency matrices, and have designed the system using inputs from social scientists.Henry and Fekete have used participatory design techniques to arrive at MatrixExplorer. Some relevant results in the form of requirements obtained from the users, who were social science researchers, are: higher preference for matrix-based representations over node-link diagrams as it was faster to display and easier to manipulate for larger data sets; cluster detection was essential for social networks analysis; and aggregating networks using clusters presented results well. To address the shortcoming of the MatrixExplorer of having huge cognitive load during context switches, Henry and Fekete have further proposed a hybrid representation, which integrates matrix and graph visualizations, called MatLink [20]. Energy-based clustering of graphs with non-uniform degrees (Lin-Log) algorithm [40] has been used for the node-link representation in MatLink.

Elmqvist et al. [9] have proposed an interactive visualization tool called Zoomable Adjacency Matrix Explorer (ZAME), for exploring large scale graphs. ZAME, which is based on representation of the adjacency matrix, can explore data at several levels of detail. In ZAME, aggregation of nodes is performed based on "symbolic data analysis" and aggregates are arranged into a pyramid hierarchy. Henry et al. have proposed NodeTrix [21] which uses the focus+context technique, where matrix visualization is used for subnetworks while graph layout is used for overview, thus confirming to the locally dense and globally sparse property. ZAME and NodeTrix use nested views [25] without using clustering; whereas our proposed representations use clustering and can be used to obtain juxtaposed views.

While these techniques explore data in the attribute space, our work focuses on exploring the same data in the similarity space. Using relationship-based approach by working in a suitable similarity space, as opposed to the high-dimensional attribute space, has been shown to be more effective for several data mining applications [43]. Additionally, similarity space can be considered to be a superset of attribute space, as identity function on adjacency matrix can be a similarity function.

Seriation Algorithms:
Seriation algorithms are ubiquitous in terms of their applications – Liiv [32] has given an overview of seriation methods, which includes the historical evolution of the technique, various algorithms used for it, and various applications, spanning across several fields, such as, "archaeology and anthropology; cartography, graphics, and information visualization; sociology and sociometry; psychology and psychometry; ecology; biology and bioinformatics; cellular manufacturing; and operations research." Mueller et al. [36], [37] have performed a thorough analysis of vertex reordering algorithms, which is relevant to representation of clusters in matrix visualization. In [36], Mueller et al. have used pixel-level display to scale up the visualization for larger data sets. In [37], Mueller et al. have referred to visual representation of adjacency matrix as visual similarity matrices (VSM) and specified common structural features of the VSM along with their graph-based interpretations. The linked views of the matrix visualization and node-link diagram show that the specific "features" in the VSM correspond to a relational pattern between the vertices in the graph. One such feature is the blocking pattern along diagonal lines, which carry information of clustering tendencies, as showcased in VAT. Though VAT focuses on blocking patterns along the diagonal of the matrix, Mueller et al. have analysed diagonal entities on and off the main diagonals.

VAT:
Bezdek and Hathway [3] have proposed a seriation algorithm along with a visual representation to assess the layout of clusters in the matrix, known as VAT (Visual Assessment of cluster Tendency). VAT uses a grayscale image of a permuted matrix to show blocks along the main diagonal, which are representative of clusters. There have been several improvisations done to VAT owing to its complexity being $O(n^2)$. Successors of VAT relevant to our work are reVAT (revised VAT) [22], bigVAT [23] and sVAT (scalable VAT) [17]. reVAT uses profile graphs to replace the dissimilarity image thus reducing the number of computation. bigVAT and sVAT are scalable variants of VAT based on sampling representative entities to reduce the size of the data. We alternatively propose parallel

implementation of VAT (pVAT) where a parallel implementation of Borůvka's algorithm is used for finding minimum spanning trees in the graph [44]. We propose pVAT because sampling algorithms, such as bigVAT and sVAT, rely on good choices of representative data and could lead to inadvertent loss of important data.

Though results of VAT depend on the starting node for the construction of minimum spanning tree, the similarity function used and the inherent clustering in the data, VAT does not involve computation of agglomerative hierarchical data structures, which becomes computationally intensive for large data sets. VAT itself can be considered to be related to single-linkage algorithm [18] as opposed to the average-linkage algorithms used in shaded similarity matrices [45], [46], [48]. Multilevel clustering has traditionally been done on the node-link diagrams using nearest neighbor consideration, which is satisfied by VAT in our proposed algorithm for multilevel clustering owing to relation of VAT to single-linkage algorithm.

Similarity Matrix Visualizations:
Two-dimensional visualization of similarity matrices has been done before, e.g., by Eisen et al. [8]. Wishart has discussed about shaded distance matrices [48] where similarity matrices are reordered by constructing a dendrogram and reordering the dendrogram to minimize the sum of weighted similarities. Wang et al. have used (a) nearest neighbor clustering and ordering using decision tree to visualize a shaded similarity matrix in [46], and (b) a combination of conceptual clustering in machine learning, and cluster visualization in statistics and graphics whose complementary properties help in interpreting clusters better, in [45]. Strehl and Ghosh [42], [43] have proposed Clusion which constructs and reorders a similarity matrix using relationship-based approach for clustering, which is an algorithm called Opposum. Erdelyi and Abonyi [10] have worked on node similarity based visualization which is very similar to ours in the use of VAT for similarity matrix representation. However they have proposed a new similarity function and implemented dimensionality reduction to obtain a similarity matrix in a lower-dimensional embedding, on which subsequently VAT is used.

These representations use an explicit clustering method which is subsequently used for reordering similarity matrices. Our work focuses on embedding the clustering with the reordering as is done in the VAT algorithm [3].

Parallel Sets:
Inselberg et al. [24] have proposed parallel coordinates for visualizing high dimensional data in a plane. Parallel coordinates representation has been effectively used for several high-dimensional data sets. Inspired by parallel coordinates, Kosara et al. [26] have proposed parallel sets for visualizing and exploring categorical data. While parallel coordinates is for representing points in multidimensional or multi-attribute space, parallel sets can be considered for discretely viewing categories. In our work, clusters can be related to categories and

hence when visualizing a series of similarity matrices, we propose using a parallel sets-like approach to get an overview of the series.

3   Basics

Similarity matrices and the VAT algorithm, which is a seriation algorithm, are pivotal to our proposed work. We describe the definitions and properties of similarity functions, seriation algorithms, and the VAT algorithm in this section.

3.1   Similarity Functions

For an ordered set of elements, applying a similarity function on pairwise choice of elements gives an idea of similarities and dissimilarities inherently present in the data. Similarity matrices are two-way one-mode representation of similarity function values between any two elements in the set. "Two-way one-mode" implies that a single set of elements is represented in two different orders, e.g., row and column orders in a matrix [32]. Subsequent clustering on a similarity matrix gives further insight to the data and can be used for reducing the complexity of the data. Similarity matrices are normalized and symmetric.

In graphs, clustering can be done in a similar fashion using similarity matrix whose elements are the vertices of the graph. In most cases the similarity matrix is a function of the adjacency matrix, since both the matrices are two-way one-mode representations of the vertices and functions for certain forms of similarity, such as structural similarity, can transform an adjacency matrix to a similarity matrix [27]. We have used the following standard similarity functions for our experiments: identity, jaccard, dice, inverse log weighted, and cosine similarity. Identity function implies that the similarity matrix is the same as the adjacency matrix.

3.2   Seriation Algorithms

An algorithm for optimal ordering of entities in a two-way one-mode representation as a square matrix, e.g. the adjacency matrix and similarity matrix, is a permutation algorithm, also known as a seriation algorithm. Ordering a set of N vertices in a graph can be done in $N!$ different ways. Graph theory states that ordering of vertices to optimize a cost function is called minimum linear arrangement (MINLA/MinLA), which is a known NP-hard problem. Hence, one uses heuristic algorithms to solve ordering of vertices to achieve domain-specific optimization criteria. An ideal ordering algorithm should be linear or better, in terms of runtime complexity. The choice of the starting vertex is critical to most ordering algorithms. We assume node 0 as starting vertex and have used the following permutation algorithms for our experiments: VAT, reVAT, Breadth first search (BFS), Depth first search (DFS), Reverse Cuthill-Mckee (RCM) [5], Kings [35], and Modified minimum degree (MMD) [33].

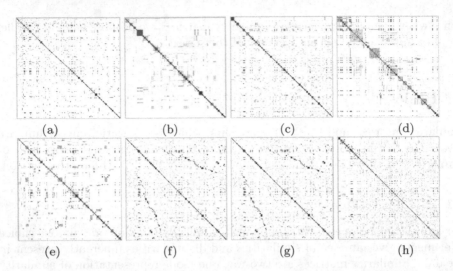

Fig. 1. Permutation or seriation algorithms showing clustering tendencies in a similarity matrix generated using cosine similarity function for a subnetwork of 372 nodes in the condensed matter collaboration network in 1999 [38]: (a) shows the unseriated similarity matrix, and (b)-(g) show the similarity matrix seriated using VAT, reVAT, BFS, DFS, RCM, Kings, and MMD algorithms, respectively. The blue highlight shows a natural cluster, which is identified by VAT but not the other seriation algorithms.

Figure 1 shows the effect of applying the permutation algorithms on the subnetwork of the condensed matter coauthorship network in 1999 [38]. For sake of clarity in showing cluster tendencies, we have chosen a subnetwork of 372 nodes from the coauthorship network which has 16725 nodes. We observe that VAT helps in identifying well-defined clusters, better than the other algorithms. As shown in Figure 1, the blue highlight shows a natural cluster in VAT, which the other seriation algorithms fail to identify clearly.

3.3   VAT

VAT [3] uses Prim's algorithm for finding the minimum spanning tree (MST) of a weighted graph to permute the order of elements in the similarity matrix. VAT shows clear clusters as black blocks along the diagonal when the matrix is represented using a grayscale image. However VAT works only on symmetric matrices and also, fails for cases where (a) there are no natural clusters in the data, and (b) the cluster types can not be identified using single linkage.

Why VAT ?:
VAT has several advantages which are apt for our application on small world networks. VAT neatly arranges clusters along the main diagonal, which helps in visual assessment as well as isolating the clusters. VAT is considered to be a single-linkage algorithm owing to its dependence on a minimum spanning tree.

This aligns with the multilevel clustering where we implicitly build a hierarchical structure.

4 Parallel Implementation of VAT

Since VAT has a runtime complexity of $O(N^2)$ for N nodes, it is not scalable for large data sets. Though bigVAT [23] and sVAT [17] are scalable, they are sampling algorithms which rely on good choices of representative data. Inappropriate choices of samples in these algorithms can result in wrong results. Hence we propose the parallel implementation of VAT (pVAT).

The serial (original) implementation of VAT uses Prim's algorithm for constructing the MST for finding the permuted order of elements in the similarity matrix. For the parallel implementation, we propose using Borůvka's algorithm for finding the MST as implemented by Vineet et al. [44] on the GPU. Borůvka's approach is generally favored in parallel implementation owing to its iterative nature. The runtime complexity of Borůvka's algorithm is $O(E \log V)$ for a graph with V nodes and E edges, while that of Prim's algorithm is $O(V^2)$, which implies that implementation of pVAT depends on both E and V, while that of original VAT depends only on V.

In VAT, we exclusively order the vertices without constructing the actual MST or evaluating the feasibility of constructing one. In a disconnected graph, a MST cannot be constructed but we can still get an ordering and construct the ordered image in VAT, which is an advantage. Both Prim's and Borůvka's algorithms give us combinatorial possibilities of MSTs. Our proposed algorithm for pVAT is as given in Algorithm 1.

4.1 Locating Clusters

The original implementation of VAT does not automatically differentiate the clusters. However for our algorithms for multilevel clustering as well as for parallel sets-like representation we require specific clusters. We use a brute-force approach as follows: in the reordered dissimilarity matrix D^* obtained using VAT, we walk along the diagonal rowwise, check columnwise the intensities of the neighboring elements to the diagonal element in the next row in D^*, and use a heuristically-derived threshold to identify start and end of clusters along the diagonal. The runtime complexity of the algorithm is $O(N^2)$ for $N \times N$ dissimilarity matrix.

5 Multilevel Clustering

In our work, we use multilevel clustering using an agglomerative hierarchical approach for achieving multiple levels of detail. It works in a bottom-up manner, initialized by all data objects as leaf nodes. For implementing multilevel clustering, we merge all the nodes in a cluster into a single node, when moving

---

Algorithm 1. pVAT: Parallel implementation of VAT

Input : $D - N \times N$ dissimilarity matrix
Output: Reordered Dissimilarity D^*
compute weighted graph G(V,E) with N vertices in V and edges in E, obtained
from using D as the adjacency matrix.
$P = \{\}$
$S = V$
for *each vertex u in V* **do**
  for *each vertex v in V, such that $v \neq u$* **do**
    $\llcorner$ find the minimum weighted edge from u to v

while *no more vertices $u \in S$ can be merged* **do**
  merge vertices $(u \in U \subseteq S)$ to form connected components, called
  supervertices (sv_U), using minimum weighted edges
  treat supervertices as new vertices, $S = S \cup \{sv_U\} - U$

for *each vertex u in S* **do**
  get the recursive ordering $O(u)$ of the subgraph in u
  $\llcorner$ $P \leftarrow P \cup O(u)$

obtain the ordered dissimilarity matrix D^* using the ordering array P as:
$D^*_{pq} = D_{P(p),P(q)}$ for $1 \leq p, q \leq N$.

---

from a finer to a coarser level of detail, and iteratively apply clustering algorithm
on the new set of nodes. The multilevel clustering algorithm terminates when
none of the nodes can be further merged, i.e, there are no more clusters. For N
nodes $(x_1, x_2, \ldots, x_N)$, if we get k clusters $(c_1, c_2, \ldots, c_k)$ after applying similarity function and seriation algorithm, where $k < N$, then the k clusters become
the nodes for the subsequent level of detail. In general, multilevel clustering is
possible only if $k < N$ strictly, as $k = N$ would imply that every cluster is a
singleton, which implicitly indicates that there are no inherent clusters in the
data set.

When moving to a coarser level of detail, the merged nodes replace the constituent nodes. For sake of simplicity, we will assume a merged node to contain
one or more nodes. Thus all the nodes in the subsequent coarser level of detail are merged nodes. When moving to a coarser level of detail, the adjacency
matrix changes, and the similarity matrix needs to be recomputed. An appropriate formulation for the attributes of a merged node has to be derived from
an appropriate aggregation of the attributes of the constituent nodes from the
finer level, which is beyond the scope of this work. Hence in our current work,
a simple similarity function of finding the maximum weighted edge between the
clusters has been used. It is equivalent to the identity similarity function, i.e.,
the adjacency and similarity matrices are the same. This function reduces a $n \times n$
matrix to a $k \times k$ matrix, where n nodes have reduced to k clusters across one
level of detail.

When using VAT as the seriation algorithm, any one of the possible minimum spanning trees is used which makes the results of our multilevel clustering

combinatorial. The number of clusters for a particular combination of data set, similarity function and seriation algorithm may vary for all transitions to the coarser levels except the last one. We have observed that the last transition always gives the same result, as the multilevel clustering terminates when no further merges can occur, i.e., the last level of detail occurs when the data set cannot be further simplified.

Multilevel clustering accentuates the need for node-cluster labelling as the new aggregated nodes should be meaningful in the context of the data set. Our work, however, is currently limited to tracking the membership of a node in a hierarchical data structure generated by multilevel clustering. The rationale behind this decision is that the membership of nodes in clusters, on its own right, implicitly reveals analytical information.

6   Visualization of Similarity Matrix Series

Similarity matrix series refers to multiple similarity matrices obtained for a set of data objects, either in the form of time series, or by application of various similarity functions or permutation functions, or by generating different subspace clusterings. Membership of objects in clusters in each of these similarity matrices is a salient aspect of the series as well as of the entire data set. Tracking the cluster-membership of objects will enable us in identifying trends and patterns in the data, thus showing the evolution of the data object across the series. We propose a parallel sets-like representation for tracking cluster-membership of objects across similarity matrices. We believe representing small world networks using this technique can give us further insights.

Kosara et al. have proposed parallel sets representation [26] of categorical data. Parallel sets has the following features: (a) it shares the parallel coordinate layout, treating all dimensions to be independent of each other, and (b) instead of using a line connecting values of coordinates to represent a data point, boxes are used to display categories and parallelograms or bands between the axes to show the relations between categories.

In our representation, the axes indicate the "instances" in the series, e.g. time-stamps in time series, one of the similarity functions in the series obtained by applying various functions, etc. The axes show the permuted order of objects in the seriated similarity matrix of the corresponding instance. We use segments in the axes to indicate clusters and lines between axes to link locations of an object in the permuted order in the different instances, similar to boxes and parallelograms in the parallel sets representation. Hence we refer to our technique as "parallel sets-like." For a series of similarity matrices, each similarity matrix can be considered to be in its own independent space, which justifies using linearly independent axes to represent them. As shown in Figure 2, each of the parallel axes corresponds to a $N \times N$ similarity matrix in the series and displays the ordered set of N data objects or nodes. Each node in the matrix has a label which can be tracked. Segments in the axes are assigned colors to indicate start and end of clusters. We use a two-color scheme where the colors are assigned alternately along each of the parallel axes to indicate start and end of clusters.

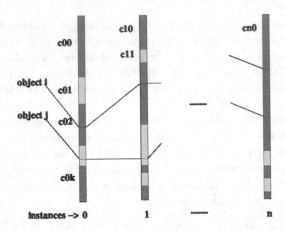

Fig. 2. Schematic of parallel sets-like representation of similarity matrices series. The parallel axes indicate the different instances $(0, 1, \ldots, n)$, which correspond to the similarity matrices in the series. The similarity matrices represented here have been permuted using a seriation algorithm, e.g. VAT, and clusters have been identified in each matrix. For each instance a, the axis is segmented using a two-color scheme to show start and end of clusters; c_{ab} indicates the bth cluster in the seriated similarity matrix a. The lines representing objects start from an initial order in the left of the axes and connect points on the axes corresponding to the object in the permuted order on the specific axis, as shown by the blue and black lines indicating objects i and j, respectively.

Generating such series of similarity matrices for a small world network and representing the series using our proposed method can aid in data mining.

7   Experiments and Results

In this section, we have described the experiments we conducted to show the effectiveness of our techniques and have shown the results for the same. We have performed case studies to show the insights we drew from using our techniques.

pVAT:
For evaluating pVAT, we have used the following data sets: coauthorship network [34], snapshots of peer-to-peer file sharing network [28], [29], who-votes-on-whom network data [31], and cocitation network [30]. Implementation of pVAT algorithm requires CUDA [41] and CUDPP [16] libraries. We have also used the igraph library [7] for graph layouts and implementing similarity functions. We have performed the experiments using Nvidia GeForce GTX 480 and 4GB RAM.

Performance measurements comparing pVAT to the serial implementation of VAT on different data sets, after applying jaccard similarity function, are as given in Table 1, which shows as expected that the former is more efficient than the latter. It is to be noted that serial implementation of VAT involves the Prim's algorithm whereas pVAT is based on Borůvka's algorithm. Hence,

Table 1. Performance measurements of serial and parallel implementations of VAT for various network data sets, after applying the jaccard similarity function

Data set	#Nodes	#Links	Serial VAT (msec)	pVAT (msec)
Coauthorship network [34]	475	1250	1×10^3	2.42
Peer-to-peer file sharing network [28]	6301	20777	1.8×10^6	4.55
Who-votes-on-whom network [31]	7115	103689	2.4×10^6	5.21
Peer-to-peer file sharing network [29]	8114	26013	3.8×10^6	4.07
Cocitation network [30]	9877	51971	6.6×10^6	3.92

Table 2. Simplification of network data sets using multilevel clustering based on VAT algorithm for seriation and cluster identification after Iteratively applying identity function on the adjacency matrix

Data set	#Nodes	#Nodes in Level Transitions	#Levels
Karate club[49]	22	22 →10→ 5→4	3
Taro exchange[14]	34	34 →7→2	2
Coauthorship[39]	1589	1589 → 615 →238 →122 →113	5
Wikipedia voting[31]	7115	7115 →3216 →1983 →927 → 725 → 635	6

we see a difference in the trends in our experiments and the speedups from the implementation in Vineet et al. [44], where serial and parallel implementations of Borůvka's algorithm are compared. Since the algorithms used for serial and parallel implementations are different, it will be incorrect to give speedups, as would be the case in a classical parallel computing scenario. Owing to the difference in dependency of the algorithms in the serial and parallel implementations on V and E, for a graph of V vertices and E edges, we observe the following from Table 1:

1. MST construction for a graph of 7K vertices and 103K edges will be faster than 8K vertices and 26K edges for the serial version, as there is no dependency of the runtime complexity on the number of edges.
2. MST construction for a graph of 7K vertices and 103K edges will be slower than a graph of 10K vertices and 52K edges for the parallel version, as there is a linear dependency on number of edges but only a logarithmic dependency on the number of vertices.

Multilevel Clustering:

For implementing multilevel clustering, we have used the following data sets: coauthorship network [39], who-votes-on-whom network data [31], taro exchange network [14], and karate club network [49]. We have used the identity function on the adjacency matrix as the similarity function and VAT as the seriation algorithm. At every transition in the level of detail, we have performed the following steps: (a) compute the new similarity matrix, (b) merge nodes in cluster to form new nodes, and (c) apply VAT on the new similarity matrix. Table 2 summarizes

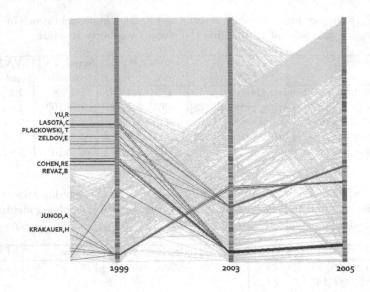

Fig. 3. Parallel sets-like representation of a subnetwork of 1036 authors taken from the condensed matter collaboration network in 1999, 2003, and 2005, to generate time series of similarity matrices. We have analyzed the clusters isolated in 1999, 2003, and 2005, highlighted by pink, blue, and red, respectively.

the number of nodes in transitions between levels of detail when implementing multilevel clustering.

Case Study 1: Time-Series of Condensed Matter Collaboration Network [38]:

In order to isolate stories of relevance in large networks, we have chosen the condensed matter collaboration networks [38] in 1999, 2003, and 2005, and created time series of similarity matrices. We have analyzed a subnetwork in condensed matter collaboration network in 1999, 2003, and 2005, which was seeded from subnetworks involving 27 authors who were common in the three data sets. We have isolated 1036 authors in total from all three instances, and applied our analysis based on parallel sets-like representation, as shown in Figure 3, where pink, blue, and red lines track authors in clusters found in 1999, 2003, and 2005, respectively.

The cluster found in 1999 shown in pink in Figure 3 comprises of the following authors: LaSota, Krakauer, Yu, and Cohen. Krakauer advised LaSota and Yu for their Ph.D. and postdoctoral research respectively circa 1999. Krakauer worked on NRL (Naval Research Lab)-funded project and Cohen worked at NRL, circa 1999. These strong proximities led to several coauthored papers amongst these four authors. In 2003, LaSota, Krakauer, and Yu continue to be in a cluster, and Cohen falls off, indicating that the connection through NRL has faded. In 2005, LaSota, Yu, and Cohen are not coauthoring with Krakauer any more, however they still form a cluster, owing to the fact that though they continue to be in

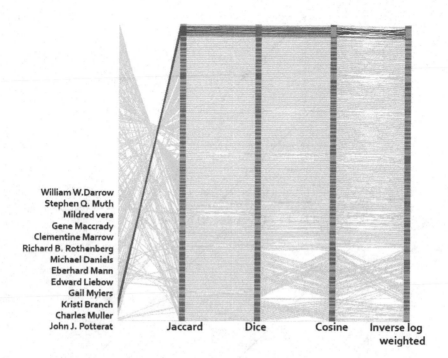

William W.Darrow
Stephen Q. Muth
Mildred vera
Gene Maccrady
Clementine Marrow
Richard B. Rothenberg
Michael Daniels
Eberhard Mann
Edward Liebow
Gail Myiers
Kristi Branch
Charles Muller
John J. Potterat   Jaccard   Dice   Cosine   Inverse log
weighted

Fig. 4. Parallel sets-like representation of a series of similarity matrices generated from applying various similarity functions and VAT in the collaboration network of social network analysts [34]. The red highlight shows the cluster that persisted through the different similarity function.

the academia in the area of physics, they are no longer active in research. The insights we derived based on coauthorship can be made directly from the graph layout, however, our techniques have additionally enabled us to gain similarity-based insights on the temporal behavioral patterns of the subjects.

The cluster in 2003 highlighted in blue consists of the following authors: Kes, Zeldov, Rappaport, Myasoedov, Feldman, Beek, Avraham, Khaykovich, Shritkman, Tamegai, and Li; who are associated with Zeldov's superconductivity lab in various capacities since 2000. Since the lab is fairly young, the associations are strong in 2003 as well as in 2005, but are scattered in 1999. The strong associations in 2003 could also have derived directly from the network owing to a paper coauthored by all of them in 2002; however the associations in 2005 is not obvious from the network directly.

The cluster in 2005 highlighted in red consists of the following authors: Wang, Junod, Bouquet, Toulemonde, Weber, Eisterer, Sheikin, Plackowski, and Revaz. They are in the same cluster in 2003, indicating that all the authors are associated through Junod by coauthorship from 2003. This inference made with the help of the parallel sets-like representation has been validated by the actual data.

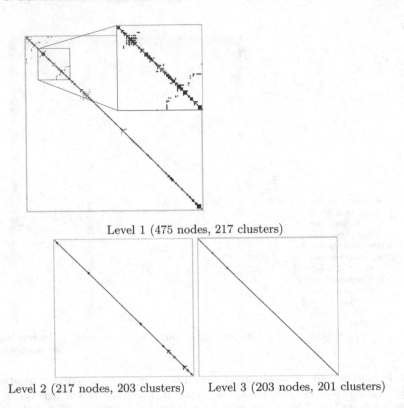

Level 1 (475 nodes, 217 clusters)

Level 2 (217 nodes, 203 clusters)   Level 3 (203 nodes, 201 clusters)

Fig. 5. Multilevel clustering of the collaboration network of social network analysts [34] using identity similarity function and VAT seriation algorithm. The blue highlight shows the nodes that merged to form clusters, which become nodes in the subsequent coarser level of detail.

Since the clusters are formed based on the cosine similarity function, we find that given a "hub", such as Zeldov and Junod, all their coauthors tend to cluster together even in the years the coauthors themselves do not publish together. This is because cosine similarity measure is based on the common neighbors and the hubs continue to act as common neighbors.

Case Study 2: Analysis of a Collaboration Network [34]:
We have applied both parallel sets-like representation and multilevel clustering on a collaboration network of social network analysts [34]. We have generated a series of similarity matrices by applying the following similarity functions: jaccard, dice, cosine similarity, and inverse log-weighted. Parallel sets-like representation of this series, as shown in Figure 4 has shown us the influence of the various similarity functions on the data set. The red highlight shows a cluster that persists through all the similarity functions. On further investigation, we find that the cluster comprises of the following authors: Darrow, Muth, Vera, Marrow, Rothenberg, Daniels, Mann, Liebow, Myers, Branch, Mueller, and

Potterat; who have co-authored with a researcher, Klovdahl, who is not in the cluster. Since the four similarity functions are based on common neighbors, the authors' association with Klovdahl causes the cluster to occur on application of each of the similarity functions.

On applying multilevel clustering on iterative application of identity similarity function and VAT, 475 nodes reduced to 217, 203, and 201, in subsequent levels of detail, as shown in Figure 5.

8   Conclusions

There have been several improvements to VAT since its inception to handle scalability. However since its scalable variants, namely bigVAT [23] and sVAT [17], are sampling algorithms, we have proposed a parallel implementation of VAT, pVAT, using CUDA and Borůvka's algorithm. pVAT enables preserving all data. As expected, performance of pVAT is more efficient compared to the serial implementation VAT which makes our tool scalable when using VAT as a seriation algorithm. The original VAT algorithm is not intended for network data, hence, no analysis of the algorithm has been done for various kinds of networks. Given the "globally sparse, locally dense" nature of small world networks, and runtime complexity of Borůvka's algorithm being $O(E \log V)$, for a graph of V vertices and E edges, we can conclude that VAT based on Borůvka's algorithm is scalable for our specific application of small world networks.

We have proposed using multilevel clustering on the similarity matrix as a way of reducing large data sets. We have found that irrespective of the variations in the results obtained from VAT, the network attains the same coarsest level of detail. Multilevel clustering, upon applying appropriate aggregation function, will enable meaningful grouping of data to achieve various levels of detail.

We have proposed using a parallel sets-like representation to visually explore multidimensional data which can be posed as a series of similarity matrices. This representation enables us to track the membership of data objects in clusters, that are identified across different similarity matrices, which correspond for different time stamps, dimensions or attributes, characteristic function, such as similarity function or seriation algorithm. Parallel sets-like representation has helped us to find significant events in time series data and interesting behavior in series derived from applying various similarity functions.

Since our work heavily depends on VAT, it comes with its limitations as well. Our future work will involve adapting our methods to address the following shortcomings: (a) representation of dense graphs as well as networks without any inherent clusters, and (b) theoretical or empirical evaluation of our assessment of the clustering capability of the similarity functions. For highly dense networks with large number of nodes where node-link diagram suffers from occlusion, matrix visualization is apt. However highly dense graphs will inherently have fewer number of clusters which will cause multilevel clustering to fail. Hence we will have to adapt our methods to work effectively on dense networks, e.g. VAT may have to be substituted by a suitable permutation algorithm for such graphs.

Though matrix representation is limited by a spatial complexity of $O(N^2)$ and is not scalable with data size, it can be resolved using pixel-level displays. Our current work does not rigorously check if VAT gives the correct number of clusters as can be obtained when applying a similarity function on the data. In order to allow for such an evaluation, we will have to formulate appropriate metrics based on our method.

Acknowledgements. The authors would like to thank Srujana Merugu, Yedendra Shrinivasan, and Vijay Natarajan for giving valuable feedback on this work. The authors would like to thank IIIT-Bangalore for their support.

References

1. Auber, D.: Tulip: A huge graph visualisation framework. P. Mutzel and M. Junger (2003), http://hal.archives-ouvertes.fr/hal-00307626
2. Auber, D., Chiricota, Y., Jourdan, F., Melançon, G.: Multiscale Visualization of Small World Networks. In: Proceedings of the Ninth Annual IEEE Conference on Information Visualization, INFOVIS 2003, pp. 75–81. IEEE Computer Society, Washington, DC (2003)
3. Bezdek, J.C., Hathaway, R.J.: VAT: A Tool for Visual Assessment of (Cluster) Tendency. In: Proceedings of the 2002 International Joint Conference on Neural Networks, IJCNN 2002, vol. 3, pp. 2225–2230. IEEE Press, Piscataway (2002)
4. Boccaletti, S., Latora, V., Moreno, Y., Chavez, M., Hwang, D.U.: Complex networks: Structure and dynamics. Physics Reports 424(4-5), 175–308 (2006)
5. Chan, W., George, A.: A linear time implementation of the reverse cuthill-mckee algorithm. BIT Numerical Mathematics 20(1), 8–14 (1980)
6. Correa, C.D., Crnovrsanin, T., Ma, K.L.: Visual Reasoning about Social Networks Using Centrality Sensitivity. IEEE Transactions on Visualization and Computer Graphics 18(1), 106–120 (2012)
7. Csardi, G., Nepusz, T.: The igraph software package for complex network research. Inter. Journal Complex Systems 1695 (2006), http://igraph.sf.net
8. Eisen, M.B., Spellman, P.T., Brown, P.O., Botstein, D.: Cluster Analysis and Display of Genome-wide Expression Patterns. Proceedings of the National Academy of Sciences of the USA 95, 14863–14868 (1998)
9. Elmqvist, N., Do, T.N., Goodell, H., Henry, N., Fekete, J.D.: ZAME: Interactive Large-Scale Graph Visualization. In: IEEE Press (ed.) IEEE Pacific Visualization Symposium 2008, pp. 215–222. IEEE, Kyoto (2008), http://hal.inria.fr/inria-00273796
10. Erdélyi, M., Abonyi, J.: Node Similarity-based Graph Clustering and Visualization, p. 483–494. Citeseer (2006), http://citeseerx.ist.psu.edu/viewdoc/download?doi=10.1.1.102.7859&rep=rep1&type=pdf
11. Ghoniem, M., Fekete, J.D., Castagliola, P.: A Comparison of the Readability of Graphs Using Node-Link and Matrix-Based Representations. In: IEEE Symposium on Information Visualization, INFOVIS 2004, pp. 17–24 (2004)
12. Ghoniem, M., Fekete, J.D., Castagliola, P.: On the Readability of Graphs using Node-link and Matrix-based Representations: A Controlled Experiment and Statistical Analysis. Information Visualization 4(2), 114–135 (2005)
13. Guare, J.: Six Degrees of Separation: A Play. Vintage Books (1990)

14. Hage, P., Harary, F.: Data set of Taro exchange (1983), http://moreno.ss.uci.edu/taro.dat
15. van Ham, F., van Wijk, J.J.: Interactive Visualization of Small World Graphs. In: Proceedings of the IEEE Symposium on Information Visualization, INFOVIS 2004, pp. 199–206. IEEE Computer Society, Washington, DC (2004)
16. Harris, M., Sengupta, S., Owens, J., Zhang, Y., Davidson, A., Satish, N.: CUDPP (2007), http://gpgpu.org/developer/cudpp-1-0a
17. Hathaway, R.J., Bezdek, J.C., Huband, J.M.: Scalable Visual Assessment of Cluster Tendency for Large Data Sets. Pattern Recogn. 39(7), 1315–1324 (2006)
18. Havens, T.C., Bezdek, J.C., Keller, J.M., Popescu, M., Huband, J.M.: Is vat really single linkage in disguise? Annals of Mathematics and Artificial Intelligence 55(3-4), 237–251 (2009)
19. Henry, N., Fekete, J.D.: MatrixExplorer: A Dual-Representation System to Explore Social Networks. IEEE Trans. Vis. Comput. Graph. 12(5), 677–684 (2006)
20. Henry, N., Fekete, J.-D.: MatLink: Enhanced Matrix Visualization for Analyzing Social Networks. In: Baranauskas, C., Abascal, J., Barbosa, S.D.J. (eds.) INTER-ACT 2007. LNCS, vol. 4663, pp. 288–302. Springer, Heidelberg (2007)
21. Henry, N., Fekete, J.D., McGuffin, M.J.: NodeTrix: A Hybrid Visualization of Social Networks. IEEE Transactions on Visualization and Computer Graphics 13(6), 1302–1309 (2007)
22. Huband, J.M., Bezdek, J., Hathaway, R.: Revised Visual Assessment of (Cluster) Tendency (reVAT). In: North American Fuzzy Information Processing Society (NAFIPS), pp. 101–104. IEEE Press, Banff (2004)
23. Huband, J.M., Bezdek, J.C., Hathaway, R.J.: bigVAT: Visual Assessment of Cluster Tendency for Large Data Sets. Pattern Recogn. 38(11), 1875–1886 (2005)
24. Inselberg, A., Dimsdale, B.: Parallel coordinates: a tool for visualizing multi-dimensional geometry. In: Proceedings of the 1st Conference on Visualization 1990, VIS 1990, pp. 361–378. IEEE Computer Society Press, Los Alamitos (1990)
25. Javed, W., Elmqvist, N.: Exploring the Design Space of Composite Visualization. In: Proceedings of the IEEE Pacific Symposium on Visualization, pp. 1–8 (2012)
26. Kosara, R., Bendix, F., Hauser, H.: Parallel sets: Interactive exploration and visual analysis of categorical data. IEEE Transactions on Visualization and Computer Graphics 12(4), 558–568 (2006)
27. Leicht, E., Holme, P., Newman, M.: Vertex similarity in networks. Physical Review E 73(2), 026120 (2006)
28. Leskovec, J.: Gnutella peer-to-peer file sharing network (August 8, 2002), http://snap.stanford.edu/data/p2p-Gnutella08.html
29. Leskovec, J.: Gnutella peer-to-peer file sharing network (August 9, 2002), http://snap.stanford.edu/data/p2p-Gnutella09.html
30. Leskovec, J.: High-energy physics citation network, January 1993-April 2003 (2003), http://snap.stanford.edu/data/cit-HepPh.html
31. Leskovec, J.: Wikipedia vote network (2008), http://snap.stanford.edu/data/wiki-Vote.html
32. Liiv, I.: Seriation and Matrix Reordering Methods: An Historical Overview. Stat. Anal. Data Min. 3(2), 70–91 (2010)
33. Liu, J.W.H.: Modification of the minimum-degree algorithm by multiple elimination. ACM Trans. Math. Softw. 11(2), 141–153 (1985)
34. McCarty, C.: Data set of network of coauthorships in the Social Networks Journal in 2008 (2008), http://moreno.ss.uci.edu/data.html#auth
35. Mueller, C.: Sparse matrix reordering algorithms for cluster identification (2004)

36. Mueller, C., Martin, B., Lumsdaine, A.: A Comparison of Vertex Ordering Algorithms for Large Graph Visualization. In: Asia-Pacific Symposium on Visualization, pp. 141–148 (February 2007)
37. Mueller, C., Martin, B., Lumsdaine, A.: Interpreting Large Visual Similarity Matrices. In: Asia-Pacific Symposium on Visualization. pp. 149–152 (February 2007)
38. Newman, M.E.J.: The structure of scientific collaboration networks. Proceedings of the National Academy of Sciences 98(2), 404–409 (2001), http://www-personal.umich.edu/~mejn/netdata/
39. Newman, M.E.J.: Data set of coauthorship network in network theory and science (2006), http://moreno.ss.uci.edu/netsci.dat
40. Noack, A.: Energy-based Clustering of Graphs with Nonuniform Degrees. In: Healy, P., Nikolov, N.S. (eds.) GD 2005. LNCS, vol. 3843, pp. 309–320. Springer, Heidelberg (2006)
41. Nvidia: CUDA (2013), https://developer.nvidia.com/category/zone/cuda-zone
42. Strehl, A., Ghosh, J.: Relationship-based visualization of high-dimensional data clusters. In: Proc. Workshop on Visual Data Mining (KDD 2001), pp. 90–99. ACM, San Francisco (2001)
43. Strehl, A., Ghosh, J.: Relationship-based clustering and visualization for high-dimensional data mining. Informs J. on Computing 15(2), 208–230 (2003)
44. Vineet, V., Harish, P., Patidar, S., Narayanan, P.J.: Fast Minimum Spanning Tree for Large Graphs on the GPU. In: Proceedings of the Conference on High Performance Graphics, HPG 2009, pp. 167–171. ACM, New York (2009)
45. Wang, J., Yu, B., Gasser, L.: Classification Visualization with Shaded Similarity Matrix. Tech. rep., GSLIS, University of Illinois at Urbana-Champaign, 9 pages (2002)
46. Wang, J., Yu, B., Gasser, L.: Concept Tree Based Clustering Visualization with Shaded Similarity Matrices. In: Proceedings of the 2002 IEEE International Conference on Data Mining, ICDM 2002, pp. 697–701. IEEE Computer Society, Washington, DC (2002)
47. Watts, D.J., Strogatz, S.H.: Collective Dynamics of 'Small-world' Networks. Nature 393(6684), 440–442 (1998)
48. Wishart, D.: ClustanGraphics3: Interactive graphics for cluster analysis. In: Classification in the Information Age, pp. 268–275. Springer (1999)
49. Zachary, W.: Data set of a university karate club (1977), http://moreno.ss.uci.edu/zachary.dat

Demonstrator of a Tourist Recommendation System

Erol Elisabeth, Richard Nock, and Fred Célimène

CEREGMIA - Centre d'Etude et de Recherche en Economie, Gestion, Modélisation
et Informatique Appliquée,
Campus de Schoelcher, B.P. 7209 97275 Schoelcher Cédex, Martinique
erol@beepway.com,
{Richard.Nock,Fred.Celimene}@martinique.univ-ag.fr
http://www.ceregmia.eu

Abstract. This paper proposes a way of using data collected from tracking gps installed in rental tourist cars. Data has been collected during more than one year. The gps positions are lined to the gps positions of the tourist sites (restaurants, beaches, museums ...). [9] These links are presented as a summary of the data. This summary is used to run specific versions of machine learning algorithms because of their geo-graphical dimension. This experiment shows how gps summaries of data can be used to extract relationships between stops of a car and touristic places.

Keywords: gps, association rules, sequential patterns, k-means, Q patterns, Geographical Center of Sequential patterns.

1 Introduction

In this paper, we begin with data summaries and we use 5 types of data mining algorithms to process these summaries: association rules, sequential patterns, Q patterns, geographical center of sequential patterns and k-Means.

The aim is to provide the best recommendation for another tourist site for a tourist in his car.

The device used in the car is a tracking gps with a PND (Personal Navigation Device). With this PND it is possible to send in real time a recommendation to the tourist, and if he accepts the recommendation, the system shows him/her the best way to join this place as any gps navigation system.

2 Data Summaries

In our demonstration we have 852 dierent data summaries. These are the activities of 12 cars during more than 14 months in 2008/2009. A path is a succession of stops between the first stop of the first day in the car and the return of the car to the park. The rows are a gps position (date, time, latitude, longitude, instant speed, altitude, cap, status of the car [stopped, running]). These rows are

V. Bhatnagar and S. Srinivasa (Eds.): BDA 2013, LNCS 8302, pp. 171–175, 2013.

used to produce a table of gps stops. After resuming, the positions collected are organized in sequential rows [date, gps stop position of, duration]. We associate the stop with the gps position of the tourist place. The result is a synthetic table with the succession of the visited tourist sites. Each row is presented as follow :

[sequential number] {date/time, tourist site, duration of visit}

[1] {date/time,1 'Casino Bateliere Plazza', 1:55}

{date/time,2 'Casino Bateliere Pointe du Bout', 3:15}

[2] {date/time,1 'Habitation Clement', 0:53}

{date/time,1 'Casino Bateliere Pointe du Bout', 5:15}

3   Association Rules

Association rule mining [2] is defined as : $I = i_1, i_2, ..., i_n$ as a set of n binary attributes called items. $D = t_1, t_2, ..., t_n$ a set of transactions called the database. Each transaction in D has a unique transaction ID and contains a subset of the items in I. A rule is defined as an implication of the form $X \Rightarrow Y$ where $X, Y \subseteq I$ and $X \cap Y = \varnothing$. The conviction of a rule [3] is defined $conv(X \Rightarrow Y)$ With association rules algorithm, it is possible to know the relationship between 2 sites. In the screen display below we can see 33,3% of tourists who went to Casino Batelire Plazza went also to Casino de laa Pointe du Bout ; but 23,1% of tourists who went to 2 : Casino de la Pointe du Bout went also to 1 : Casino Batelire Plazza. With the proposed map we can see that the behaviors of tourists from south of the island are dierent from the one from the center.

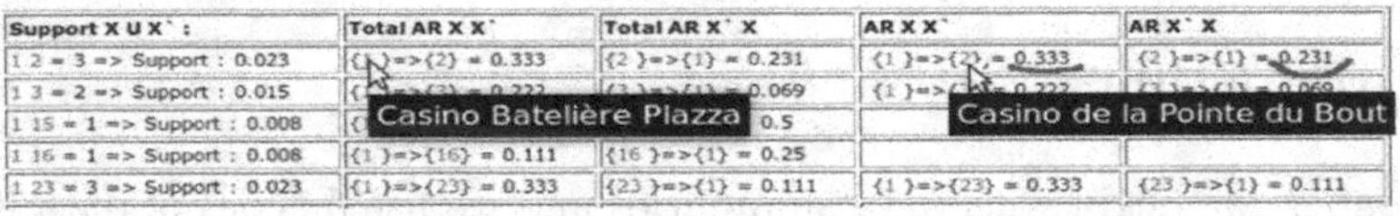

Support X U X` :	Total AR X X`	Total AR X` X	AR X X`	AR X` X
1 2 = 3 => Support : 0.023	{1 }=>{2} = 0.333	{2 }=>{1} = 0.231	{1 }=>{2} = 0.333	{2 }=>{1} = 0.231
1 3 = 2 => Support : 0.015	{1 }=>{3} = 0.222	{3 }=>{1} = 0.069	{1 }=>{3} = 0.222	{3 }=>{1} = 0.069
1 15 = 1 => Support : 0.008	Casino Batelière Plazza	0.5	Casino de la Pointe du Bout	
1 16 = 1 => Support : 0.008	{1 }=>{16} = 0.111	{16 }=>{1} = 0.25		
1 23 = 3 => Support : 0.023	{1 }=>{23} = 0.333	{23 }=>{1} = 0.111	{1 }=>{23} = 0.333	{23 }=>{1} = 0.111

Fig. 1. Association rule between 2 sites

4   Sequential Patterns

Extraction of sequential patterns [1] and [4] make possible the discovery of temporal relations between 2 sites. In this sequence we notice a relationship between 52 : 'ART et Nature' AND 257 : 'Hotel le Panoramique'. We may suppose that before coming back to the 'hotel Le Panoramique' the tourist went to 'Art et Nature'. We propose two types of representations. The geographical map of pattern allows to have a visual representation of behaviors of tourist stops. It is possible to create an oriented diagram that shows the inter site links.

(52) ART et NATURE	=>	(257) Hotel Le Panoramique **	9	0.0124826629681
(1) Casino Batelière Plazza	=>	(23) BIBLIOTHEQUE SCHOELCHER	6	0.0083217531207
(1) Casino Batelière Plazza	=>	(3) La Galleria	2	0.0027739251 0402
(1) Casino Batelière Plazza	=>	(80) HERTZ	4	0.0055478502 0604
(1) Casino Batelière Plazza	=>	(2) Casino de la Pointe du Bout	2	0.0027739251 0402
(188) TOTAL	=>	(257) Hotel Le Panoramique **	2	0.0027739251 0402
(2) Casino de la Pointe du Bout	=>	(80) HERTZ	4	0.0055478502 0804
(2) Casino de la Pointe du Bout	=>	(134) Anse Noire	2	0.0027739251 0402
(2) Casino de la Pointe du Bout	=>	(1) Casino Batelière Plazza	2	0.0027739251 0402
(2) Casino de la Pointe du Bout	=>	(267) Hotel Pagerie ***	2	0.0027739251 0402
(2) Casino de la Pointe du Bout	=>	(137) Anse Mitan	2	0.0027739251 0402
(3) La Galleria	=>	(80) HERTZ	7	0.0097087378 6408
(3) La Galleria	=>	(23) BIBLIOTHEQUE SCHOELCHER	7	0.0097087378 6408
(3) La Galleria	=>	(280) Hotel Valmeniere***	5	0.0069348127 6006

Fig. 2. sequential pattern between sites

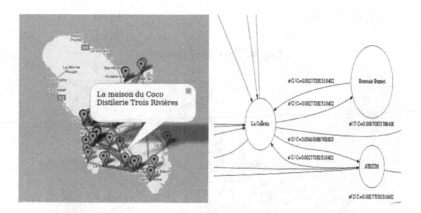

Fig. 3. Map of sequential pattern Diagram inter site links

4.1   Q Patterns

The Q patterns are like patterns but the item sets do not have a fixed dimension [5] and [8]. For example item sets are in the database $conv(A, B \Rightarrow C)$ and in the same database we can also have $conv(A, B, D, F \Rightarrow T)$

23 : Bibliothque Schoecher , 217 : La Kasa Saveurs, 298 : Karibea Rsidence La Goelette, 12 : Habitation Clment, 39 : Habitation Depaz, 40 : Distillerie Neisson, 130 : Grande Anse d Arlet

$(23, 217, 298, 12, 298, 39, 40, 298 \Rightarrow 130)$ We can have a representation where each pattern has a specific color on an oriented graph. This sequential pattern shows a succession of stops with the same topic : rum distillery. We can have a representation where each pattern has a specific color on an oriented graph. Each Bubble is a tourist site.

4.2   Geographical Center of Sequential Patterns

We also compute a geographical solution as a recommendation when we already have found the sequential pattern or the cluster in a specific k-means. The objective is to find -in real time- the best tourist site next to a car when we already

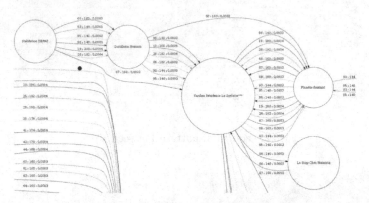

Fig. 4. Part of Q pattern

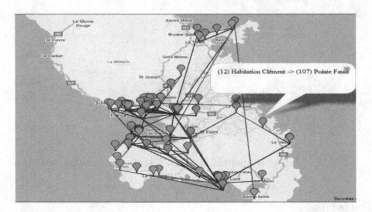

Fig. 5. Map of centroid of sequential patterns

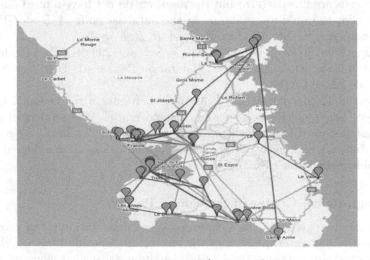

Fig. 6. K means (3 clusters)

classify it in a k-means cluster or a sequential pattern [6] and [7]. In this example if the car is IN the cluster (12, 107) or in a sequential pattern where there are item sets (12 AND 107) and if the car is next to the centroid of this data, we can propose a new activity.

4.3   k-Means

We can have a k-means representation using 2 clusters. The cluster A (in red) is around the Center of the island and the south, the second one B (in yellow) is in the south.

5   Conclusion

In this paper, we have applied a set of data mining algorithms using data collected from tracking gps installed in rental tourist cars. Tourist organizations and agencies could look into these applications to find the best way to extract knowledge from their own database systems. GPS tracking companies can also find ideas to improve the uses of their collected data.

References

1. Agrawal, R., Srikant, R.: Mining sequential patterns. In: Yu, P.S., Chen, A.L.P. (eds.) Proceedings of the Eleventh International Conference on Data Engineering, March 6-10 (1995)
2. Mannila, H., Toivonen, H.: Multiple uses of frequent sets and condensed representations (extended abstract) (1996)
3. Lallich, S., Teytaud, O.: Evaluation et validation de l'interet des regles d'association (2000)
4. Nicolas., P., Lotfi, L.: Data mining: Algorithmes d'extraction et de rduction des regles d'association dans les bases de donnes (2000)
5. Zaki, M.J.: Spade: An efficient algorithm for mining frequent sequences. Machine Learning 42(1/2), 31–60 (2001)
6. Masseglia, F., Teisseire, M., Poncelet, P.: Extraction de motifs sequentiels. Problemes et methodes (2004)
7. Pei, J., Han, J., Mortazavi-Asl, B., Wang, J., Pinto, H., Chen, Q., Dayal, U., Hsu, M.-C.: Mining sequential patterns by pattern-growth: The prefixspan approach. IEEE Transactions on Knowledge and Data Engineering (2004)
8. Grozavu, N., Bennani, Y.: Classification collaborative non supervisee LPN UMP CNRS, 249-264 CAP (2010)
9. Béchet, N., Aufaure, M.-A., Lechevallier, Y.: Construction et de structures hirarchiques de concepts dans le domaine du e-tourisme: INRIA - 475-506, IFIA (2011)

Performance Comparison of Hadoop Based Tools with Commercial ETL Tools – A Case Study

Sumit Misra[1], Sanjoy Kumar Saha[2], and Chandan Mazumdar[2]

[1] RS Software (India) Ltd.
[2] Jadavpur University, Department of Computer Science & Engineering

Abstract. Data analysis is one of the essential business needs of organizations to optimize performance. The data is loaded into data warehouse (DWH) using Extract, Transform and Load (ETL). Analytics is run on the DWH. The largest cost and execution time is associated with the ET part of this workflow. Recent approaches based on Hadoop, an open source Apache framework for data intensive scalable computing, provide an alternative for ET which is both cheaper and faster than commercial prevalent ETL tools. This paper presents a case study where experimental metric results have been presented in support of the claim. The reduction of cost makes it viable for small and large organizations alike and reduction in execution time makes it possible to provide online data services.

Keywords: ETL, Big Data, Hadoop, Cost, Performance.

1 Introduction

The volume of organizational data is increasing at a very high rate [1]. This data has diverse sources, formats, quality and frequency of generation. The data needs to be collected, cleaned, curated and stored in a way that information retrieval and analysis for business intelligence becomes easy. This demand is accentuated due to requirement of collapsing processing window duration. ETL constitutes 60%-80% [2] of business intelligence projects out of which ET is the major component.

The usual approaches to address this issue have been to add hardware, or adopt a faster ETL tool, or reduce refresh cycle of master data, and similar measures. However, now the data volume has reached to such a level where these steps are either increasing operational cost or reducing responsiveness of the business, or both, hence non-sufficient. This has lead us to evaluate options for moving from high performing ETL tools to new technology open source low cost option that uses Map Reduce (M/R) paradigm [3,4]. We have implemented such solutions and the results are promising – as it can reduce cost and improve performance. Newer tools are emerging that uses the M/R paradigm which promises even faster processing and can take streaming inputs as well.

While conducting survey of similar work we find reference which highlight absence of such data due to the license cost of commercial ETL tools [5]. We have

V. Bhatnagar and S. Srinivasa (Eds.): BDA 2013, LNCS 8302, pp. 176–184, 2013.
© Springer International Publishing Switzerland 2013

relied on both our own experiments and experience of other researchers to arrive at a comparison of performance, infrastructure cost and resource cost. The findings will be useful for the business community to take appropriate decisions.

Rest of the paper is organized into 3 sections. Section 2 outlines the Case Study that highlights the possible outcome of the study and compares the solution architectures. Section 3 details the Experiment and Result. Section 4 is the Conclusion which summarizes the findings and its impact.

2   Case Study

The goal of this case study is to present a comparison on cost and performance between commercial ETL and open source Map Reduce (M/R) based solutions. A dual approach of surveying and conducting experiments has been adopted. The scope of the work is limited to the extract and transform (ET) part of the ETL effort which is reported as the major component of business intelligence service workflow.

2.1   Outcome

The study would help to assess the applicability of the business solutions options based on following:

Processing Speed.
As the ETL constitutes 60%-80% of the Business Intelligence (BI) and Data Warehouse projects, enhancement of the performance of ETL enables the opportunity to provide near-real-time data services.

Cost.
Where proprietary ETL cost ranging from $50K to $200K per license, Hadoop, an open source Apache project built using M/R, offers low cost distributed processing using cluster of commodity machines connected using gigabit LAN. Using such low cost processing frameworks, it makes the solution affordable to even small and medium business houses thus reducing the barriers to entry.

Resource Development.
Proprietary ETL tools are usually not available at affordable prices even for building resource talents. Thus the available skill base for such ETL architects, designers, developers and testers is very low which in turn increases the cost as premium amount needs to be paid for rare skills. As open source tools facilitates low cost alternative it would be easy to develop resources with requisite skill base. Thereby the total cost of ownership (TCO) for the business organizations gets reduced.

2.2   Comparison of the Solution Architectures

The primary challenge of ETL is to handle very large data sets. The ETL tools that implemented parallelism in their architecture addressed the challenge using three forms of parallelism [2] as depicted in Fig 1.

- Data Parallelism – where the data is split into small segments and processes can work on each of these segments in parallel.
- Component Parallelism – where multiple instance of same component can process different segments of task in parallel
- Pipeline Parallelism – where the data segments are processed in sequence by series of processing task components such that when data segment[i] is under process by task component [k], simultaneously data segment [i+1] is under process by task component [k-1].

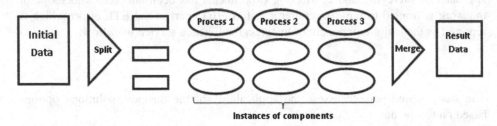

Fig. 1. Schematic diagram of implementation of parallelism in ETL

Advantage of the above architecture is that it scales well if the hardware is scaled out in the servers running the ETL. The disadvantages are as under

- Usually the data is not local to the processing node that results in underuse of local processing power and overuse of network resources
- If there is a fault in the hardware the system cannot recover gracefully
- The cost of scaling up is very high as server components needs to be scaled out, for example, processor needs to be upgraded from dual core to quad core, etc
- Some of the ETL licenses are core based, hence when one upgrades the processor the licenses need to be scaled up suitably.

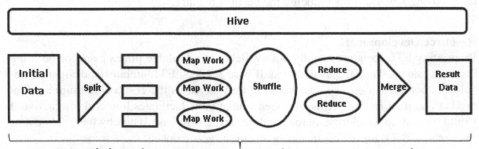

Fig. 2. Schematic diagram of M/R which is used by Hive

In M/R framework [6], the underlying processing is done in two sequential steps, Map followed by Reduce. Within both the phases' considerable parallelism can be exploited. Between Map and Reduce there is a phase called Shuffle which is executed by the M/R framework. In this phase all the output from Map phase processes having same key is grouped. Each key along with its grouped data set is allocated to Reduce phase for further processing. This simple construct is very powerful and abstractions such as Hive converts a SQL-like query to a M/R construct which gets executed as a M/R job. The architecture is represented in Fig 2.

Advantage of M/R architecture is that it can run on commodity hardware and can be easily scaled out by adding such machines. It does not require scaling up of the components of the machines, thus addressing the scalability issue is economic. In general, the workers (map, reduce) are replicated at least 3 times and a master node (i.e. Name Node) assigns the tasks to these worker nodes and monitors their progress. Hence, if one of the workers fails, master node can detect the same and use the results of the other nodes for continued processing. Thus the design is highly fault tolerant. Thus the advantages of the Hadoop based architecture can be summarized as

- Easily scalable architecture
- Scaling up is economic
- Design is highly fault tolerant

The disadvantage is limited to the need for using gigabit LAN connectivity between the machines for faster transmission of larger blocks of data to worker machines.

3   Experiment and Result

In this section we would describe the data profile and process flow used for the experiment. This will be followed by the results.

Data Profile.
Behaviors of the data fields present in the file are quite different. Some of them are affected by almost every transaction whereas some are static. Domain of certain fields may be large whereas some are flags with value coming from very small list. Profile of the various fields in the data file is presented in Table 1.

Table 1. Profile of the fields of data used in the experiment

Field 1	It represents the users account number. There is hardly any repetitions as there are 99.88% distinct values. Hence it is concluded that most of the transactions are from distinct account holders.
Field 2	This field represents where the user purchased from, and there seems to be a limited set of shops which contributed to the data as the variation is only 0.65%
Field 3	This indicates the category of the shop. There are about 1000 possible variations of the categories and the data covers almost 500 categories. This indicats that the coverage of the categories was good (75%)

Field 4	This indicats the country of the shop. The data covered about 96.9% of the possible variations.
Field 5	This is a flag and can take only 2 values
Field 6	This is a small set with only 10 distinct values and all were covered
Field 7	This indicated the fees involved and resulted in coverage of about 90% of the possible variations

Number of records used for this experiment was in the order of 1500 million.

Process Flow.
The experiment of ETL is built around the task of finding the aggregates for transaction data grouped by certain categories. A standard ETL requirement was conceived that necessitated the following tasks to be conducted on a set of transaction data:

- Extract the data from a flat file with comma separated value (CSV)
- Remove duplicates
- Transform the data by joining a lookup table and replacing codes (keys) by the corresponding values
- Roll up the data on a set of dimensions and generate an output file

As it has been mentioned that we have focused on the ET aspects, loading the data was not part of the experiment. Fig 3 depicts the processing flow of the ETL.

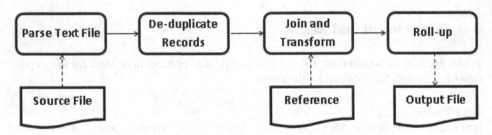

Fig. 3. Schematic diagram the ETL process for the experiment

3.1   Experiment I

The first experiment was conducted on production like equipment and a large data set. The objective was to provide an estimate of the performance and cost of using a commercial ETL infrastructure as compared to using M/R based open source infrastructure. [1]

The infrastructure used for the experiment id detailed in Table 2.

---

[1] Certain exact details could not be shared due to confidentiality reasons but that does not impact on the information required to assess the cost or performance.

Table 2. Comparison of the infrastructure used for ETL and M/R cluster

	Setup A (ETL)	Setup B (M/R based)
Tool Used	Proprietary ETL	M/R based Hive
Hardware Used	Large server with 3+ GHz 8-12 core processor 128-256 GB RAM; ETL was assigned, 8-way multithreaded processing	Cluster of 8-24 commodity scale machines connected by 1 GBPS LAN running Hive on top of the cluster. The master node (name node) was 2.5 GHz 8-16 processor 2-core with 48-72 GB RAM blade server. The commodity machines have 2+ GHz 2-core processor with 8 GB RAM and 500 GB HDD
Data Volume	400 GB	400 GB

Performance.
Throughput obtained for the two setups is depicted in the Table 3. This was carried out in private data centers. It clearly indicates the improvement achieved for the Apache Hadoop based system.

Table 3. Comparison of the throughput for ETL and M/R cluster

	Setup A (ETL)	Setup B (M/R based)	Time saving for Setup B
Throughput	39.45 seconds / GB, or 91.25 GB / hour	9.75 seconds / GB, or 369.23 GB / hour	75.28%

Hence there was **75.28% saving in time** when M/R is used instead of the ETL.

Infrastructure Cost.
Cost of the software for performing the data processing involved a server side and a client side component. The client side component was essentially the integrated development environment (IDE) to build the logic for the flow. The execution was entirely done using the server side component. The cost of the server side component for commercial tools depends on the number of CPU and number of core per CPU. The client side was purely based on number of users using the IDE.

Cost comparison for the two setups is shown in Table 4. License cost which is payable annually is primarily used for assessing the infrastructural cost. Thus for a 10 member team running 2-core CPU servers, the license cost saving would be in the order of **77.78%.** Typically 15%-20% of hardware cost is annual maintenance cost for the hardware. Considering hardware cost and its maintenance, the saving in infrastructure TCO is **about 75% empirically**.

Table 4. Comparison of the infrastructure cost in terms of licenses and hardware requirements

	Setup A (ETL)	Setup B (M/R based)	Cost saving for Setup B
Software License	Server: $100K to $250K per CPU Per User: $10K [7]	Server: $0 Per User: $0 to $10K [8]	Assuming 10 member 2-core server CPU saving is 77.78%
For 2-core CPU and 10 member team max license cost	Average ($100K, 250K)*2 + $10K*10 = $450K	10K * 10 = $100K	

Manpower Cost .

Table 5. Comparison of the ratio of the salary along with ease of developing skill-set

	Setup A (ETL)	Setup B (M/R based)	Manpower cost saving for Setup B
Salary Ratio [9]	100%	77%-96% (Average 86.50%)	About 13.50%
Ability to train	Difficult [10]	Easy	

There could be resource **cost saving of about 13.50%** assuming uniform spread of resources used for Setup B as shown in Table 5.

The result of Experiment I can be summarized in Table 6.

Table 6. Improvement in metrics for M/R based tool with respect to standard ETL tools

#	Parameters	Change
1	Performance improvement	75.28%
2	Infrastructure TCO reduction	75.00%
3	Manpower cost reduction	13.50%

3.2   Experiment II

The objective of this experiment was to eliminate the impacts of network infrastructure and the impact of multiple nodes in the cluster and review how the tools leveraged multi-threading and data access efficiently. This was accomplished by using a single server and run the ETL and various M/R based tools on a smaller subset of the data on similar server environment and implemented the same workflow as in the case of the previous experiment.

Data Volume used was 220MB. The server used was 3GHz, 2-Core processor with 4 GB RAM and 500 GB HDD.

Table 7 provides the performance of the ET of the data for the various toolsets. It may be noted that the performance data is obtained after discounting network overheads. Elapsed time stands for the processing time of the data. As it had been already mentioned, we had used single server in order to eliminate network delays and overheads.

Table 7. Performance comparison of M/R based tools and ETL when run on one machine

#	Tool used	Elapsed time in seconds	Percentage of ETL
1	M/R on Apache Hadoop Framework	152	20%
2	Pig on Apache Hadoop Framework	186	24%
3	Hive Query on Apache Hadoop Framework	233	30%
4	Commercial ETL	765	100%

It is evident that in terms of speed 70%-80% gain is achieved in case of Hadoop framework over commercial ETL tools.

From the experiments I and II it is observed that the technology of MR on Apache Hadoop framework provides better performance as MR exploits distributed processing.

4   Conclusion

Analyzing the past works and carrying out the experiments we conclude that Hadoop based solutions are better in comparison to commercial ETL tools. Experiment I has indicated there is substantial improvement in terms of throughput, reduction in infrastructure and manpower cost. Experiment II clearly shows how M/R is taking advantage of the threading and data access to produce a better yield. This highlights the clear advantage of moving from commercial ETL tool to Map Reduce based solutions. The commercial houses are adopting Map Reduce into their solution architecture, however, the TCO remains quite high and this is promoting adoption of Map Reduce based solutions. The companies like Talend and Syncsort who were in lower segment as per ability to execute [11] are now moving up and right to the desired quadrant. In future we would have to carry out the experiments with multiple data sets of bigger sizes to study the applicability of the technology on a wide variety of data sets.

References

1. Halevi, G., Moed, H.: The Evolution of Big Data as a Research and Scientific Topic. Research Trends 30 (2012)
2. Eckerson, W., White, C.: Evaluating ETL and Data Integration Platforms. Report of The Data Warehousing Institute (2003)

3. Ferguson, M.: Offloading and Accelerating Data Warehouse ETL Processing using Hadoop. Report of Intelligent Business Strategies (2013)
4. Rodriguez, N., Lawson, K., Molina, E., Gutierrez, J.: Data Warehousing Tool Evaluation – ETL Focused. In: Proc. SWDSI 2012 (2012)
5. Liu, X., Thomsen, C., Pedersen, T.B.: ETLMR: A Highly Scalable Dimensional ETL Framework Based on Map Reduce. In: Cuzzocrea, A., Dayal, U. (eds.) DaWaK 2011. LNCS, vol. 6862, pp. 96–111. Springer, Heidelberg (2011)
6. MapReduce Overview, Google App Engine, https://developers.google.com/appengine/docs/python/dataprocessing/ (accessed July 2013)
7. Cost of ETL tools, http://enselsoftware.blogspot.in/2009/06/cost-of-etl-tools.html
8. 5 Common Questions About Apache Hadoop Accessed (July 2013), http://blog.cloudera.com/blog/2009/05/5-common-questions-about-hadoop/
9. Map Reduce Salary, http://www.indeed.com/ (accessed July 2013)
10. ETL Tools Comparison, http://www.etltools.net/etl-tools-comparison.html (accessed July 2013)
11. Friedman, T., Beyer, M.A., Thoo, E.: Magic Quadrant for Data Integration Tools, Gartner Report (2010)

Pattern Recognition in Large-Scale Data Sets: Application in Integrated Circuit Manufacturing

Choudur K. Lakshminarayan[1] and Michael I. Baron[2]

[1] Hewlett Packard Research, USA
Choudur.Lakshminarayan@iith.ac.in
[2] University of Texas at Dallas, USA
mbaron@utdallas.edu

Abstract. It is important in semiconductor manufacturing to identify probable root causes, given a signature. The signature is a vector of electrical test parameters measured on a wafer. Linear discriminant analysis and artificial neural networks are used to classify a signature of test electrical measurements of a failed chip to one of several pre-determined root cause categories. An optimal decision rule that assigns a new incoming signature of a chip to a particular root cause category is employed such that the probability of misclassification is minimized. The problem of classifying patterns with missing data, outliers, collinearity, and non-normality are also addressed. The selected similarity metric in linear discriminant analysis, and the network topology, used in neural networks, result in a small number of misclassifications. An alternative classification scheme is based on the locations of failed chips on a wafer and their spatial dependence. In this case, we model the joint distribution of chips by a Markov random field, estimate its canonical parameters and use them as inputs for the artificial neural network that also classifies the patterns by matching them to the probable root causes.

Keywords: Integrated-Circuit, Failure Analysis, Signature Analysis, Pattern Recognition, Linear Discriminant Analysis, Mahalanobis Distance, Neural Networks, Back Propagation, Markov Random Fields and spatial signatures.

1 Introduction

SIGNATURE analysis (SA) is a statistical pattern recognition program designed to assign failed parts to one of several pre-determined root cause categories. Engineers invest lots of time tracing back test probe/electrical parameter failures to probable root causes. It is desired to have an automated program based on sound statistical theory that enables the classification of a failing signature to a root cause category such that the probability of misclassification is minimized. Linear discriminant analysis (LDA) is an established parametric procedure that minimizes the probability of misclassification and allows the failure analysis engineer to state "The probability that a failing wafer with a specific signature belongs to the kth root cause category is p %."

Signature analysis is intended as an aid and not a replacement for sound engineering analysis and judgment. It is not meant for situations where the root cause is quite

V. Bhatnagar and S. Srinivasa (Eds.): BDA 2013, LNCS 8302, pp. 185–196, 2013.

obvious, for it does not contribute anything that the engineer does not already know. It is really intended for the more subtle situation where the root cause of the failure is not readily apparent. Instead of, or in addition to the signatures, one can classify wafers and assign them to known root causes based on the locations of defective chips on a wafer. Analysis of production wafers indicates that patterns of failing clusters, their direction, size, and shape differ from one root cause to another and hence they can be used as inputs to the classification scheme. Information on spatial location of failed chips is summarized in ten parameters of a suitable Markov random field and used as input layer node elements for an artificial neural network to classify wafer patterns into root cause categories.

This paper is organized in the following manner. In section 2 we introduce linear discriminant analysis as a methodology for classification of patterns to root causes, decision rules that maximize the probability of correct classification, and the assumptions required for its implementation. Section 3 deals with issues related to missing data, outliers, collinearity within the features of an input pattern, and non-normality. In section 4 we present the artificial neural network approach to pattern classification. In section 5 we present some examples of the implementation of linear discriminant analysis and artificial neural networks in IC failure analysis. Section 6 develops a classification scheme based on locations of failed chips and the general condition of an anisotropic Markov random field. In section 7 we draw conclusions.

2   Linear Discriminant Analysis and Mahalanobis Distance

Assume there are k root cause categories, and let j denote the jth category where $j = 1,2,3,...,k$. Let p be the number of electrical parameter measurements defining each signature denoted by z; thus a signature can be expressed as a $1 \times p$ row vector. Let n_k be the number of signatures in the kth root cause category. Let μ_k denote the mean vector (centroid) of dimension $1 \times p$. The elements of μ_k are the averages of the p test parameters defining the signature. Let Σ be the common covariance matrix, whose elements are variances and covariances of the p electrical test parameters. It is also assumed that the signature constitutes a sample random vector from a *multivariate normal distribution*. A multivariate normal distribution is a p-dimensional extension of the one variable Gaussian distribution. The decision rule that maximizes the probability of correct classification under the assumption of multivariate normality of an incoming failing signature $z = (z_1, z_2,...,z_p)$ into root cause category j is given by $argmax_j d_j(x)$, where

$$d_j(x) = ln(p_j) - (1/2)(z-\mu_j)^T \Sigma^1 (z-\mu_j), \quad j = 1,2,......,k.$$

An incoming signature vector z is classified into category iif $d_j(x) > d_i(x)$ for $j \neq i$. Equivalently, vector z is classified into category C_j for which D_j is minimum. In practice, μ_j and Σ are unknown, and are estimated from the sample. The sample estimates of the mean vector and the covariance matrix are given as $\overline{x}$ and S. The application

LDA requires *equicovariance*; which corresponds to the property of equal covariance structure across the k root cause categories. Combining the sample covariance matrices results in a pooled covariance matrix, S_p (*pooled sample covariance*). An estimate of the Mahalanobis distance metric using pooled covariance is given by:

$$\hat{D}_j = \left(z - x_j\right)^T S_p^{-1}\left(z - x_j\right)$$

References [1, 2] provide a detailed discussion of discriminant rules and Bayes classifiers. There are no definite guidelines for the number of sample signatures in each root cause category, but it is recommended that each n_j be at least ten times larger than p. The asymptotic chi-square property of the Mahalanobis distance statistic allows one to associate a probability of misclassification or its complement. A new incoming failing pattern z will be classified in category j if $\hat{D}_j$ is the minimum over all $j=1,2,...,k$. The Mahalanobis distance alone is used for the relative ranking among the root cause categories. A small value of $\hat{D}_k$ in conjunction with a large probability of correct classification suggests that the pattern does belong to the root cause k.

3   Implementation

The actual computation of the Mahalanobis distance for a p-variate pattern vector is a simple matter. However, implementation of an automated pattern recognition program in failure analysis (FA) is an involved proposition. Several methodological as well as logistical issues need to be addressed to successfully execute a pattern recognition program. In the following sub-sections we discuss the source of electrical test data, data formats, filtering data, missing data, and constructing root cause databases etc., in elaborate detail.

3.1   Sources of Electrical Test Data

Test data is obtained from a Keithley parametric tester. The Keithley is a parametric tester that obtains measurements on electrical test parameters. The tester is interfaced to a prober that is used to probe the process control bars on a wafer that contain test structures to measure the desired parameters. Typical test structures are resistors, capacitors, and transistors. Electrical contact to the structures is established via bond pads. When contact is established by the probes to the test structures, computer programs instruct the tester to measure the desired test structure characteristics. This data is analyzed to determine the device parameters on the process control bars on a wafer, prior to testing, visual inspection for superficial defects, and on to packaging for shipping to the end user.

3.2   Data Format

The resultant data from a Keithley tester is formatted in a columnar form for readability by the pattern recognition program. The data stream consists of lot number, wafer number, site number, and the measurements corresponding to the electrical parameters. A p-dimensional row vector of electrical test measurements constitutes a signature. A typical signature is given by $x = \left(x_1, x_2, ..., x_p \right)^T$. We will refer to a signature also as a pattern, and the constituents of a signature as features. Typical electrical test parameters are: drive current, breakdown voltage, sheet resistance, P+SD resistance, and N+SD resistance etc. It is typical in IC failure diagnostics that electrical test parameters in excess of 300 are studied to capture anomalies. Layout of the architecture of the SA program is given in Figure 1. Each root cause set consists of signatures and the corresponding root causes including the lot number, the wafer number, and the site number.

3.3   Databases

It is evident from Figure 1 that databases of root causes are a key to the implementation of the SA program. A file consisting of lot number, wafer numbers, site numbers, vector of electrical test measurements, and root causes is compiled. The file is separated by root causes and the k individual root cause databases is established. Upon building the root cause databases, after working through the issues of test data, the root cause specific statistics such as mean vectors and covariance matrices are computed which are the ingredients that go to compute the Mahalanobis distance metric.

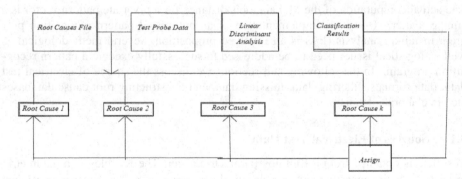

Fig. 1. Signature Analysis Program Layout

3.4   Handling Missing Data

Missing data occurring sporadically is not a problem and is readily handled. The term "sporadic" means that no more than five percent of the data is missing for any given parameter for each root cause set. The missing data is replaced by the corresponding component sample mean from the mean vector for that root cause. This substitution induces certain *artificialness* in the data, but enables the computation of the covariance matrix. Missing data could correspond to sparsely populated features in a signature, and also features that are absent. Eliminating sparse features (>5% missing) in a given root cause could lead to signatures of unequal dimension residing in different root cause categories. If this is the case, the largest intersection of the components of the two vectors is taken. For example, sample mean vector $\overline{x}_1$ for root cause set 1 consists of components $(x_1, -, x_3, x_4, x_5)$ and the mean vector $\overline{x}_2$ for root cause 2 consists of components $(x_1, x_2, -, x_4, x_5)$, then the intersection of these two four dimensional vectors (four features present) would be the three dimensional vector with components $(x_1, -, -, x_4, x_5)$. The greatest intersection method allows measurements of the same quantities to be compared. The greatest intersection is employed across all k root cause categories. For this reason, the root cause categories are carefully chosen so they are of the same dimension in the signatures inhabiting them. For example, if there are seven root cause categories where the dimension of the signatures are all the same, say 60 features, then the addition of an eighth root cause category with only fifty features would eliminate the information from 10 electrical test parameters. It should be noted that signatures are technology dependent since each technology consists of different test parameters. It is therefore essential to build databases of signatures and root causes by technology for SA implementation.

3.5   Collinearities and Outliers

The dimension of the signatures should be reduced whenever possible to alleviate cumbersome and often unnecessary calculations. The device analysis engineer should eliminate those parameters that duplicate information. Always, judicious reductions in the dimension of the signature should be constantly sought to avoid collinearities. Also, we employ the method of *step-wise discriminant analysis* to find an optimal subset of features in a signature. See [3] for a brief overview of step-wise discriminant analysis.

Classification by linear discriminant analysis is sensitive to *outliers*. Outliers have harmful effects on classification. In the present application, any measurement beyond $\pm 4.5\sigma$ (standard deviations) away from each component mean is deemed an outlier. Using this outlier window, harmful effects of outliers are mitigated.

3.6   Effects of Non-normality

Electrical test parameter measurements do not satisfy the assumption of multivariate normality. Severe departures from normality have a large effect on LDA. Since the LDA statistic follows aChi-square distribution, any departure from the assumed statistical distribution results in unreliable probability estimates.

4   Artificial Neural Networks

An artificial neural network (ANN) is an arrangement of neurons organized in layers. A *multi-layered* feed-forward network consists of an input layer of neurons, a hidden layer of neurons feeding into an output layer of neurons. A schematic of a fully connected, multi-layered feed-forward neural network with one hidden layer is illustrated in Figure 3. The ANN is trained using the back propagation algorithm [4].

4.1   Training with Back Propagation

A multi layered feed-forward neural network with one hidden layer was chosen for training with the following characteristics:

• The number of elements in the input signature was chosen to be 34 after a careful review of significant parameters for the technology under investigation.

• The data is scaled such that the domain of any given electrical parameter is the unit interval (0,1).

• Sufficient training data relative to fourteen known root causes was established prior to training.

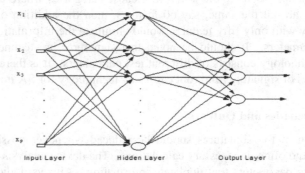

Fig. 2. Architecture of a multi-layer feed forward neural network

• The number of hidden layer units was chosen to be a third of the sum of input and output layer units. The number of hidden layer units is recommended by [4]. The computation amounts to 16 hidden layer units.

• The hidden layer and output layer weights (synaptic strengths) were generated from a Gaussian distribution with mean 0 and variance 0.3. It is recommended a small value for the variance is chosen.

• All the bias units were set to zero in the entire network.

• The fourteen output classes were set to an identity matrix of dimension 14x14 the largest element (1) replaced by 0.9 and the smallest element (0) replaced by 0.1.

• The training data was split into three sets, namely the training set, validation set, and the test set as recommended by [6].

• The training set was used to train several preliminary network architectures

• The validation set was used to identify the network with the least sums of squares of error.

• The third test set was used to assess the performance of the chosen network with examples independent of data used in training.

• The network was trained using the pattern by pattern approach. The patterns were selected randomly from each root cause category to eliminate any biases during learning.

4.2   Comments

To extract best classification performance, we needed to remove correlations within the signature, and achieve data parsimony. *Principal components analysis* (PCA) is a viable approach to enable both objectives [10]. In addition to eliminating correlations among the features of the signature vector, PCA enables reduction in the dimension of the signature vector while retaining bulk of the information contained in the original vector. We urge the user to consider PCA for the dual purpose of signature vector de-correlation and dimensionality reduction.

5   Results

The following are some examples of the application of LDA and artificial neural networks for integrated-circuit failure analysis. To validate the two techniques, we saved some signatures with known root causes for testing. The saved signatures did not contribute to the sample statistics computed or participate in neural network training.

Example. Two lots failing due to missing N+S/D implant were submitted to the automated signature analysis program for root cause identification. A signature of length 34 is applied to the program for pattern classification. The signature is from a certain device XXXXXXXXX belonging to technology YY. Tables 1 and 2 show the results of this analysis. The number of electrical test parameters measured for this technology is 123, but a signature of dimension 34 is applied for classification. The reason is that a careful selection of 72 test parameters for SA was chosen by engineering. Further, application of step-wise discriminant analysis reduced the dimension to 34.

The program classified signatures from each site into a root cause category. It is known that site 2 on wafer 17 was failing due to missed N+ S/D implant problem. LDA clearly identified the failing signature to belong to the known root cause. The probability column corresponds to the probability that the signature is from a population known to possess N+ S/D implant problem as the root cause. The low D value with high probability of classification suggests that the algorithm was able to accurately classify the new signature. The output summarizes the statistics relative to the top four probable root causes. Applying the neural network (standard and PCA) approach to classification, the results are consistent with those from LDA. Classification by a neural net is realized by choosing that root cause for which the error sum of squares is small. We reiterate that the error sum of squares is the square of the accumulated error between an input signature and the desired response. In the second example form lot 9749493, wafer 3, and site 2, although the distance metric D due to LDA was the smallest for the known root cause category, the probability of classification was very small.

Table 1. Classification by Linear Discriminant Analysis and Artificial Neural Networks

Lot Number	9745158
Device	XXXXXXXXX
technology	YY
Number of Wafers	24
Number of Sites	5
Number of Parameters	123

LDA by Site				
Wafer Number	Site Number	MD	Root Cause	probability
17	2	9.65	Missed N+S/D implant	0.980000
17	2	2.03.84	Missed Nwell Implant	0.000000
17	2	367.48	High Epi Doping	0.000000
17	2	408.77	Sidewall Overetch	0.000000

Lot Number	9745158
Device	XXXXXXXXX
technology	YY
Number of Wafers	24
Number of Sites	5
Number of Parameters	123

ANN by Site			
Wafer Number	Site Number	squarred error	Root Cause
17	2	0.4857	Missed N+S/D implant
17	2	2.7752	Missed Nwell Implant
17	2	2.884	Missed DUF Implant
17	2	2.9477	Via

Lot Number	9745158
Device	XXXXXXXXX
technology	YY
Number of Wafers	24
Number of Sites	5
Number of Parameters	123

ANN (principal components) by Site			
Wafer Number	Site Number	squarred error	Root Cause
17	2	0.6221	Missed N+S/D implant
17	2	0.8144	High Epi Doping
17	2	0.9815	Missed DUF Implant
17	2	1.0953	Via

Table 2. Classification by Linear Discriminant Analysis and Artificial Neural Networks

Lot Number	9749593
Device	XXXXXXXXX
technology	YY
Number of Wafers	24
Number of Sites	5
Number of Parameters	123

LDA by Site				
Wafer Number	Site Number	MD	Root Cause	probability
3	2	1333.1	Missed N+S/D implant	0.000000
3	2	1666.11	Thin Gate Oxide	0.000000
3	2	1817.46	Missed Nwell Implant	0.000000
3	2	1849.86	Low Epi Doping	0.000000

Lot Number	9749593
Device	XXXXXXXXX
technology	YY
Number of Wafers	24
Number of Sites	5
Number of Parameters	123

ANN by Site			
Wafer Number	Site Number	squarred error	Root Cause
3	2	0.2937	Missed N+S/D implant
3	2	1.619	Thin Gate Oxide
3	2	1.7003	Missed Nwell Implant
3	2	1.7286	Low Epi Doping

Lot Number	9745158
Device	XXXXXXXXX
technology	YY
Number of Wafers	24
Number of Sites	5
Number of Parameters	123

LDA by Site				
Wafer Number	Site Number	MD	Root Cause	probability
17	2	9.65	Missed N+S/D implant	0.980000
17	2	2.03.84	Missed Nwell Implant	0.000000
17	2	367.48	High Epi Doping	0.000000
17	2	408.77	Sidewall Overetch	0.000000

The computation of the probability of classification hinges on the distribution of sample D which is asymptotically chi-square under the assumption of multivariate normality of the signature vector. It is our suspicion that the statistical distribution of the signature vector is not multivariate normal. The data is perhaps from a heavy tailed distribution causing the probabilities to be small. The neural network found the appropriate root cause with the smallest error sum of squares.

6   Classification Based on Locations of Failed Chips

In addition to signatures, one can use the information of spatial patterns of failed chips to classify wafers into known root cause categories. Recent studies [12, 13, 14] attempted to develop a sensible stochastic model that would reflect the complicated spatial dependence structure of failing chips. The anisotropic *Markov random field* model proposed in [12] extends a simpler model of [13], allowing different strengths and different directions of the spatial correlation of failures and capturing most of the significant patterns of clusters of failing chips. Goodness-of-fit tests show strong significance of this extension. A training sample of 388 wafers produced the following results of chi-squared goodness-of-fit tests.

Table 3. Results of the chi-squared goodness-of-fit tests

Significance level	0.5%	1%	2.5%	5%	10%
Percentage of rejected hypothesis	93.3%	95.1%	96.9%	97.7%	98.2%

For example, if additional parameters modeling anisotropic spatial patterns were not significant, the 10% level test would have rejected the null hypothesis with probability 0.10. However, it actually rejected the hypothesis for 98.2% of training wafers.

The assumption of a Markov random field means that any two chips are dependent on a wafer, however, failure of one chip, conditioned on the surrounding chips, occurs independently of the rest of the wafer. Mathematically this means

$$P\{chip\ (i,j)\ fails \mid the\ rest\ of\ the\ wafer\} = P\{chip\ (i,j)\ fails \mid neighborhood\ of\ (i,j)\},$$

where (i,j) denotes x- and y-coordinates of a chip on a wafer, and the neighborhood of (i,j) consists of all the chips $(k,m) \neq (i,j)$ satisfying $|k-i| \leq 1$ and $|m-j| \leq 1$.

Theoretical development of this stochastic model in [12] shows that the joint distribution of all the chips on a wafer is a function of 10 canonical parameters. They represent the probability for a chip to fail regardless of other chips (pure failure rate, or main effect), the probability to fail because of a failure of its neighbor (pairwise interactions), the probability to fail given simultaneous failure of two of its neighbors (second-order interactions), and the probability to fail given simultaneous failure of three neighbors (third-order interaction). In order to estimate these parameters, one defines $X_{km}=1$ if the chip (k,m) is good and $X_{km}=0$ if it is defective and computes ten local joint-count statistics, $N_\alpha=X_{km}$, $N_\beta=\Sigma X_{ij}X_{i,j+1}$, $N_\gamma=\Sigma X_{ij}X_{i+1,j}$, $N_\delta=\Sigma X_{ij}X_{i-1,j+1}$, $N_\varepsilon=\Sigma X_{ij}X_{i+1,j+1}$, $N_\eta=\Sigma X_{ij}X_{i,j+1}X_{i+1,j}$, $N_\zeta=\Sigma X_{ij}X_{i-1,j}X_{i,j+1}$, $N_\iota=\Sigma X_{ij}X_{i,j-1}X_{i-1,j}$, $N_\kappa=\Sigma X_{ij}X_{i,j-1}X_{i+1,j}$, and $N_\lambda=\Sigma X_{ij}X_{i,j+1}X_{i+1,j}X_{i+1,j+1}$. Each sum Σ is taken over all combinations of cliques that include the chip (k,m). Then, one computes conditional distributions of X_{km}, given its neighborhood Q_{km}. For every defective chip, it equals

$$P\{X_{km} \mid Q_{km}\} = (1+e^{-<\Theta,N>})^{-1},$$

where $\Theta=(\alpha,...,\lambda)$ is the vector of unknown parameters and $N=(N_\alpha,...,N_\lambda)$ is the vector of joint-count statistics. For every good chip (k,m), this conditional distribution equals

$$P\{X_{km} \mid Q_{km}\} = (1+e^{<\Theta,M>})^{-1},$$

where M is the vector of joint-count statistics obtained from the same wafer, with the chip (k,m) replaced by a defective chip. The pseudo-likelihood function $L(\Theta)$ is then defined as the product of conditional probabilities for all the chips on a wafer. Finally, canonical parameters $\alpha,...,\lambda$ are estimated by the ten-dimensional vector that maximizes $L(\Theta)$ in Θ. This can be done, e.g. by means of the Matlab function *fminu*. Obtained maximum pseudo-likelihood estimates carry information on the most typical patterns of clusters of defective chips on a wafer, their typical shape, size, width, direction. Goodness-of-fit tests show significant improvement of the earlier models based on N_α, N_β, and N_γ only.

Also, ten estimated parameters serve as inputs for an artificial neural network that classifies wafers according to spatial patterns and matches to known root causes. Out of the total 780 training patterns, 675 were classified correctly, based on the Markov random field only (instead of signatures) and the back-propagation algorithm, described in Section 4, yielding a classification rate of 87%.

7   Conclusion

Linear discriminant analysis and artificial neural networks are viable tools for pattern classification in an integrated-circuit manufacturing environment. Depending on the type of data collected from wafers, a practitioner chooses a method based either on signatures (Sections 2-5) or on spatial locations of chips on a wafer (Section 6). The Markov random field model is applicable to produced wafers only whereas the methods based on LDA and ANN can be implemented during the course of production. For the implementation of these methods, it is paramount to build a database that consists of a plethora of root causes. This helps to suitably associate signatures to probable root causes. We implemented ANN architectures wherein the nodes in every layer were fully connected. It may be advantageous to investigate topologies that are sparsely connected but result in improved performance. Currently we are studying alternate ANN and artificial intelligence models for classification in IC failure diagnostics. We considered modeling die electrical signatures using logistic regression methods for pattern classification, and it turns out that discriminant analysis and artificial neural networks lend themselves useful implementation in the manufacturing line.

References

1. Johnson, R.A., Wichern, D.W.: Applied Multivariate Statistical Analysis, 3rd edn. Prentice Hall, Englewood Cliffs (1992)
2. Duda, R.O., Hart, P.E.: Pattern Classification and Scene Analysis. John Wiley & Sons, New York (1973)
3. McLachlan, G.J.: Discriminant Analysis and Statistical Pattern Recognition. John Wiley & Sons, New York (1992)

4. Freeman, J.A., Skapura, D.M.: Neural Networks, Algorithms, Applications, and Programming Techniques, Computation and Neural systems Series. Addision Wesley, Reading, Massachusetts (1991)
5. Press, W.H., Teukolsky, S.A., Vetterling, W.T., Flannery, B.P.: Numerical Recipes in C, 2nd edn. Cambridge University Press, Cambridge (1996)
6. Ripley, B.D.: Pattern Recognition and Artificial Neural Networks. Cambridge University Press, Cambridge (1996)
7. Haykin, S.: Neural Networks. A Comprehensive Foundation. MacmillanCollege Publishing, New York (1994)
8. Tobin, K.W., Gleason, S.S., Karnowski, T.P., Cohen, S.L., Lakhani, F.: Automatic Classification of Spatial Signatures on Semiconductor Wafermaps, Private communication
9. Gleason, S.S., Tobin, K.W., Karnowski, T.P.: Spatial Signature Analysis of Semiconductor Defects for Manufacturing for Manufacturing Problem Diagnosis, Solid State Technology (July 1996)
10. Jolliffe, I.T.: Principal Component Analysis. Springer, New York (1986)
11. Hand, D.J.: Discrimination and Classification. John Wiley and Sons, New York (1981)
12. Baron, M., Lakshminarayan, C.K., Chen, Z.: Markov Random Fields in Pattern Recognition for Semiconductor Manufacturing. Technometrics 43, 66–72 (2001)
13. Longtin, M.D., Wein, L.M., Welsh, R.E.: Sequential Screening in Semiconductor Manufacturing, I: Exploiting Spatial Dependence. Operations Research 44, 173–195 (1996)
14. Taam, W., Hamada, M.: Detecting Spatial Effects from Factorial Experiments: An Application from Integrated-Circuit Manufacturing. Technometrics 35, 149–160
15. Snedekor, G.W., Cochran, W.G.: Statistical Methods, 8th edn. Iowa State University Press, Ames (1989)

Author Index